ENGINEERING PROCEDURES HANDBOOK

ENGINEERING PROCEDURES HANDBOOK

by

Phillip A. Cloud

np

NOYES PUBLICATIONS

Westwood, New Jersey, U.S.A.

Library of Congress Catalog Card Number: 97-26403
ISBN: 0-8155-1410-7

Published in the United States of America by
Noyes Publications
369 Fairview Avenue
Westwood New Jersey 07675

Transferred to Digital Printing, 2011

Printed and bound in the United Kingdom

Library of Congress Cataloging-in-Publication Data

Cloud, Phillip A.
Engineering procedures handbook / by Phillip A. Cloud.
p. cm.
Includes index.
ISBN 0-8155-1410-7
1. Engineering--Management. 2. Engineering--Documentation.
I. Title.
TA190.C62 1997
651.5'0426--dc21 97-26403
CIP

FOREWORD

This handbook contains 62 engineering procedures and 47 forms. Most of these engineering procedures are influenced by the author's background in aircraft, aerospace and the computer industry. The manufacture of Printed Circuit Boards, was used as an example, throughout this book. Principles discussed in this handbook, however, are applicable to all engineering and operational disciplines.

The handbook is divided into seven sections that are in the logical order of a product engineering documentation system:

Product Design and Development
Product and Document Identification
Documentation Requirements
Customer Documentation
Vendor Documentation
Document Change Control
Document Control

The intent of this handbook is to define, display, and explain the elements that must be present in any engineering documentation system. The author has developed a powerful new handbook, complete with step–by–step procedures, that will show you how to develop an engineering documentation system.

Primarily, it will be used as a reference source when there is a question as to the proper procedure to use, for example: how do you document a situation where a critical vendor will need to be approved as a single source? You would look in the Contents under Single Source and it will direct you to the procedure EP–5–2 Single Source Authorization. Most of the procedures in this handbook have a cross reference system in Section 2.0, that guides you to other related information.

ABOUT THE AUTHOR

Philip A. Cloud is a freelance writer with over forty years of industrial experience. He founded his own company in 1984, Cloud Publishing Co., specializing in hardware and software documentation with specific expertise in document development, release, change control, reproduction, and distribution. He has written over 25 administrative and technical manuals for industry.

Previously, he was Configuration Management Manager, Engineering Documentation Manager, and Drafting Manager. He has also been a Configuration Management Engineer, Technical Writer, Standards and Procedure Writer, Software Quality Engineer, Engineering Analyst, Design Engineer, Engineering Drafter, Tool Design Engineer, Validation Specialist.

He has worked with corporation such as: Geneva Pharmaceuticals, Inc., Storage Technology Corp., Miniscribe Corp., Micro Motion, IBM (contract writer), Beech Aircraft Corp., The Boeing Co., Gates–Lear Jet, and Cessna Aircraft Co.

He has personally guided the development of the engineering documentation control process at several corporations and has made significant contributions toward improving the product release process at several other companies. He has implemented Material Requirement Planning systems as well as a variety of other projects in his field.

Notice

PREFACE

Documentation is the section against which most noncompliances are written during ISO 9000 registration audits. One reason is that implementation of requirements for documentation is checked in all departments of a company. This significantly increases the risk that a documentation noncompliance will be found. Documentation is the area where companies are most vulnerable.

Engineering documentation audit failures are caused by: procedures for specific operations do not exist, procedures are outdated and not available for employee use, and there is no way of assuring that procedures are actually followed all the time.

An engineering organization that is committed to continuous improvement and cost control is unbeatable. One way of being unbeatable is having a documented engineering system. Achieving successful engineering documentation management consistently requires a documented system for planning and control. A company can save thousands of dollars by implementing an engineering documentation system.

If an individual or company wishes to create or improve as engineering documentation system, there is no need to start from scratch. Instead use this new *Engineering Procedures Handbook*, complete with 47 specially designed forms and with procedures that cover every major aspect of a comprehensive engineering documentation system. Another book published by Noyes Publications, *Engineering Documentation Control Handbook*, can be very helpful if used in conjunction with this handbook.

This handbook is a new systematic approach to engineering documentation, therefore, it will simplify the end users ability to set up or enhance their engineering documentation requirements. Companies with small manual systems, to large scale mass production facilities, can use this handbook to tailor their engineering documentation requirements.

The separate elements of this handbook when joined together, represent a total system that describes tasks, shows their relationships, and assigns responsibilities. Thus, the concept of an organization is required. For this reason, a mock organization was created for this handbook:

> President, Vice President, Administration, Marketing/Sales, Purchasing, Engineering, Manufacturing, Quality.

Organizational structures, names, and functional responsibilities vary from company to company. A first task for individuals using this handbook is to see what variations exist between this handbook and their own organizations, and make allowances and adjustments accordingly. Later, when editing the procedures themselves, care must be taken to ensure that correct names and proper responsibility assignments are used.

Phillip A. Cloud

CONTENTS

LIST OF FIGURES

SECTION 1

PRODUCT DESIGN AND DEVELOPMENT

Title: **PRODUCT PLANNING**		Number: EP-1-1 Revision: A
Prepared by:	Approved by:	

1.0 PURPOSE

The purpose of this procedure is to provide a list of some of the major items to be considered when developing a new product or a major line extension.

2.0 APPLICABLE DOCUMENTS

Engineering Procedures:

EP-1-2, Product Introduction

EP-7-1, Document Control System

3.0 INTRODUCTION

Successful and timely introduction of new products or major line extensions and product improvements assures continued growth and profits to the company.

This procedure is written as a guide to all company employees who are involved in product development and product introductions. It is designed to discipline product development activities, leading to a speedier, more cost effective product development cycle.

Through the use of checklists, master schedules and defined reviews, oversights which can lead to extra development time and increased project or costs will be eliminated.

3.1 Product Planning Process

This procedure describes the product planning system. The system provides a systematic framework for evaluating new business opportunities. This process will help accomplish:

- a consistent approach to evaluate new products
- prioritizing projects
- establish schedules
- a method to audit progress
- improve communications
- a consensus for a product or project
- minimized project expense and product cost

Decisions to continue to the next phase are made by the Product Review Board. The Product Review Board is comprised of: Engineering, Manufacturing, Quality, Finance, Marketing/Sales and others as designated by the Manager of Engineering.

3.2 Product Idea

Any person associated with the company can suggest a new product idea. In order to begin the development process, the idea originator must accomplish the following steps:

Complete the Preliminary Product Summary information.

Provide additional information as necessary to explain the product idea.

Present information to the Product Review Board for review and approval.

The Preliminary Product Summary phase must be approved by the Manager of Engineering after review with the Product Review Board in order to proceed to the Preliminary Investigation phase.

The Product Review Board will assign a project manager and marketing coordinator to complete the Preliminary Investigation phase.

3.3 Project Organization

It is intended that each new product or major line extension project be led by a project manager. The project manager is selected by the Product Review Board after a new product or major line extension has been approved for preliminary investigation.

The project manager manages the new product or major line extension project from a company standpoint. A project manager can be selected from any functional group within the company. Administratively, project managers will continue to report to their functional managers. The project manager's position is one of the important elements in timely execution of the new product development program and thus, dedicated effort, with review by company officers, is necessary for success.

Reporting on matrix bases to the project manager are coordinators from functional disciplines involved in design, development, manufacturing and testing of the new product. A typical project team will consist of coordinators from the following departments: Marketing, Customer Service, Manufacturing, Engineering, Purchasing and Accounting. Coordinators from other functional areas may be selected on a per product or as needed basis.

3.4 Coordinator Responsibilities

The following provides a list of basic responsibilities for each primary member of the project team.

3.4.1 Engineering Review Board.

Approve expenditures for major product and projects.

Assign responsibilities for direction and completion of projects.

Assure conformance of projects to long range goals and budgets.

Assure completion of projects in a timely fashion.

3.4.2 Project Manager.

Coordinate preparation of project budgets and master schedules for the project.

Coordinate preparation and release of specifications for the project.

Monitor and control the preparation, publication and tracking of project schedules.

Prepare monthly summaries of project progress for Product Review Board review presenting and discussing information when necessary.

Coordinate activities required to meet project budgets and schedules and promptly reporting any deviations or problems to the Product Review Board.

Coordinate formal design and project reviews when necessary as outlined in these procedures.

Evaluate the performance of project team members. Forward evaluations to respective department managers.

3.4.3 Engineering Coordinator.

Prepare budgets, schedules and specifications for the engineering phase of the project.

Assist in preparation of real worth analysis.

Communicate with other team members to accomplish on-line release of the product.

Assure that engineering efforts are completed on time and to specification.

3.4.4 Marketing/Sales Coordinator.

Assure that Marketing/Sales inputs and requirements are part of specifications.

Assure that trade-offs in the design of the product do not jeopardize the market.

Coordinate and prepare sales projections.

Coordinate preparation of product literature and advertising.

Coordinate sales force training on the product.

In conjunction with Engineering, Customer Service, establish field test sites and coordinate data collection.

Prepare Marketing introduction plan.

3.4.5 Manufacturing Coordinator.

Coordinate production tooling design.

Coordinate development of routings and process.

Coordinate development of cost standards.

In conjunction with Manufacturing coordinator, assure that production lines, etc. are properly set up.

Participate in the specification and acquisition of production test equipment.

Coordinate product cost estimates.

Assure that personnel and systems are in place for the pilot run and production of product.

Assure that manufacturing inputs are considered in the design phase of the project.

Participate in cost analysis of the product.

Prepare procedures necessary for production of new products.

3.4.6 Quality Coordinator.

Assure that quality and reliability considerations are part of the product design.

Assure that tooling and production test equipment meet quality requirements.

Assure that vendors have the required quality approval.

Assure that inspection equipment and procedures are in place.

Perform quality and reliability audits and tests on prototypes and pilot runs.

Perform product safety audits.

3.5 Project Reviews

The project manager will hold regular project team meetings, reporting on routine basis, via meeting minutes and project status reports. Attendance at project meetings is required of all team members.

On a monthly basis, during monthly Product Review Board meetings, time will be provided for review of major projects. The project manager will highlight product performance to specification: cost vs. plan, timetable vs. plan, competitive evaluations, major milestones for balance of project and review major project highlights and project problems.

Formal reviews will occur at the beginning of each phase of the project. Reviews will assess project and product performance against the plan. A successful review is required to proceed to the next phase of development.

3.6 Documentation

Proper documentation is critical to long-term success of new products. The requirements of an "engineering project file" are being emphasized here. Ultimately the team members are responsible for documenting their results, with the project manager responsible for the overall coordination of this effort.

The project manager will establish and maintain a project file which documents the decisions and events of concern to the project. (Reference: EP-7-1, Document Control System.)

3.7 Checklist and Forms

Checklists and forms need to be prepared by representatives of the functional groups involved in the planning, design, manufacturing, quality, reliability, safety and introduction of a new product. The intent of the checklist is to ensure all relevant details necessary for the development of a successful product are addressed. As a result, several of the checklists may overlap one another. Through disciplined use of checklists you can expect to achieve:

A comprehensive product definition.

A design that meets the product definition.

A program that stays within cost and schedule estimates.

A complete and orderly market introduction.

A documented history of the development decisions (to be used to make informed decisions to support production of the new product.)

Project to project consistency.

3.8 Schedule

A project schedule will need to be developed that will show the approximate order of the project milestones. This schedule is to be used as a guide for developing detailed schedules for new projects.

In order to obtain maximum benefits from project schedules, each item listed in it should be broken down into finer detail. Completion dates and activity duration can then be estimated with a higher probably of being accurate.

Title: **PRODUCT INTRODUCTION**		Number: EP-1-2 Revision: A
Prepared by:	Approved by:	

1.0 PURPOSE

The purpose of this procedure is to provide instructions and to assign responsibilities for assigning project numbers to new products. It will also provide instructions for preparing and submitting the Product Introduction Notice form.

2.0 APPLICABLE DOCUMENTS

Engineering Procedure:

EP-1-1, Product Planning

EP-2-2, Document Number Assignment Logbook

Engineering Form:

E001, Product Introduction Notice

3.0 OVERVIEW

3.1 New Product Introduction

For each new product introduction, a Product Introduction Notice form shall be prepared, approved and submitted to Document Control. Document Control will log, copy, stamp, distribute and file the new Product Introduction Notice form.

3.2 Project Classification Changes

For Product Classification changes a Product Introduction Notice form shall be prepared and submitted to Document Control. Document Control will log, copy, stamp, distribute and file the Product Introduction Notice form and closed out the previous project number, if applicable.

4.0 PRODUCT INTRODUCTION PROCEDURE

This procedure is divided into the following parts:

- Product Introduction Procedure
- Product Introduction Notice Form Preparation

The following procedure describes who is responsible and what they are supposed to do for each processing step.

4.1 Engineer

Steps

1. Prepare the Product Introduction Notice form E001 to assign a new classification number to a new product. (See Figure 1. for an example of the form.) Generate a new project number using the following numbering system:

4.1.1 New Project Classification Number Assignment

XXXXX - Classification

100-199 - Patent — Applies to projects created to track time spent on patent support.

200-299 - Development — A product that is currently in the Development phase.

300-399 - Production — A product that is currently in the Production phase.

400-499 - Maintenance — Maintenance of a product that is in the Production phase.

500-599 - Other — A product that does not fit any of the above descriptions.

XXXXX - Department

1 - Hardware Engineering

2 - Software Engineering

3 - Electrical Engineering

4 - Documentation only

5 - Manufacturing Support

6 - Quality Support

4.1.1 New Project Classification Number Assignment (Continued.)

XXXX<u>X</u> - Type

1 - Stand-alone	A single item.
2 - Cluster	A product made up of items.
3 - PCB - Stand-alone	A printed circuit board for a single item.
4 - PCB - Cluster	A printed circuit board for a product made up of items.
5 - Other	A product not fitting any of the above descriptions.

4.1.2 Project Number Interpretation

Following is an example of how to interpret a project number:

Example project number 20521.

205 = A product currently in the Development phase

2 = Software Engineering

1 = Stand-alone

2. Forward the completed Product Introduction Notice form to the Engineering Manager for review and approval.

4.1.3 Engineering Manager

3. Verify that the Product Introduction Notice form is complete and correct. If approved, sign and date, then return the form to the Engineer for further processing.

4.1.4 Engineer

4. Forward the completed Product Introduction Notice form to Document Control for logging, copying, stamping, distributing and filing.

4.1.5 Document Control

5. Log in the information from the new Product Introduction Notice form into the Product Introduction Number Assignment Log. (Reference EP-2-2, Document Number Assignment Logbook, under Paragraph 4.0 Product Introduction Number Assignment Log.)

6. Run copies, stamp, distribute, then file the original in the Open Projects File by project number.

4.2 Project Classification Change Number Assignment

4.2.1 Engineer

7. If the project classification changes (e.g., the project progresses from Development to Production, or from Production to Maintenance, etc.), prepare and submit a Product Introduction Notice form to Document Control for logging, copying, stamping, distributing and filing.

4.2.2 Document Control

8. Log in the information from the new Product Introduction Notice form into the Product Introduction Assignment Number Log. (Reference EP-2-2, Document Number Assignment Logbook, under Paragraph 4.0 Product Introduction Number Assignment Log.)

9. Run copies, stamp, distribute, then pull the old Product Introduction Notice form original from the Open Projects File, and file it in the Closed Projects File by project number.

PRODUCT INTRODUCTION NOTICE		
Engineer: ❶	Date:	Project No: ❷
Eng. Manager: ❸	Date:	Product: ❹
Project Title: ❺		
Project is: ❻ ❑ Current, give existing No.________________ ❑ Subset ❑ New		
Project Classification: ❼ ❑ Patent ❑ Development ❑ Production ❑ Maintainance ❑ Other, specify:		
Purpose of Project : ❽		

Form E001 (Procedure EP-1-2)

Figure 1. Product Introduction Notice Form

5.0 PRODUCT INTRODUCTION NOTICE FORM PREPARATION

The procedure for assigning new project numbers shall be followed by each individual responsible for entering information on the Product Introduction Notice form. Each circled number below corresponds to the circled number on the Product Introduction Notice form E001. (See Figure 1. for an example of the form.)

5.1 Engineer

❶ Enter your name and the date when the form is complete and ready for processing.

❷ Enter the project number using the document numbering system shown in Paragraph 4.1.1.

5.3 Engineering Manager

❸ Enter your name and the date, if approved.

5.4 Engineer

❹ Enter the product name.

❺ Enter the project title.

 Mark the appropriate box:

Project is:

Current - if requesting a new project number for an existing project, then enter the current project number.

Subset - if creating a project that is related to an existing project (e.g., a project to develop a different component of the same product).

New - to open a new project.

 Mark the appropriate box:

Project Classification:

Patent - if the project is for patent support.

Development - if the project is for a product that is in the Development phase.

Production - if the project is for a product that is in the Production phase.

Maintenance - if the project is for maintenance of a product that is currently under production.

Other - if none of the above classifications apply, specify the classification.

8 Enter the purpose of the project.

Title: **MAKE-OR-BUY COMMITTEE**		Number: EP-1-3 Revision: A
Prepared by:	Approved by:	

1.0 PURPOSE

The purpose of this procedure is to provide instructions and to assign responsibilities for the Make-or-Buy Committee members.

2.0 APPLICABLE DOCUMENTS

None.

3.0 DEFINITIONS

3.1 Make Item

An item produced or work performed by the company.

3.2 Buy Item

An item produced or work performed outside the company.

3.3 Must Make Item

An item or service which the company regularly makes and is not available (quality, quantity, delivery, and other essential factors considered) from outside vendors.

3.4 Must Buy Item

An item or service which the company does not have the capability to provide.

3.5 Can Make-or-Buy Item

An item or service that can be provided by the company or outside vendor at prices comparable to the company (considering quality, quantity, delivery, risk and other essential factors).

4.0 OVERVIEW

4.1 Make-or-Buy decisions are to be based on good business practices. The following factors are considered in arriving at Make-or-Buy decisions:

Total cost to customers.
Schedule considerations, including risks.
Quality of service or product.
Performance of item or service.
Complexity of item or difficulty of administering control.
Loading of functional organizations which would normally perform the task to assure that adequate capacity exists.
Facilities and capital equipment requirements.

Availability of competition, especially from small business firms and labor surplus area firms.
Quality of technical data package used for procurement of an item or service.
Technical and financial risks associated with potential suppliers.
Security considerations.

4.2 Make-or-Buy Committee

The Make-or-Buy Committee is established to evaluate and decide on all Make-or-Buy Plans, items and services presented to it.

The Committee consists of the following members or their designees. Voting members are indicated by a (V):

Project Manager - Chairman (V)
Project Support Analyst - Secretary (Records minutes and maintains official file.)
Engineering Manager (V)
Manufacturing Manager (V)
Quality Manager (V)
Purchasing Manager (V)
Finance Manager
Contract Administration

5.0 MAKE-OR-BUY PROCEDURE

The following procedure describes who is responsible and what they are supposed to do for each processing step.

5.1 Project Support Analyst

Examine all Request for Quotes and Request for Proposals to determine the need for a Make-or-Buy Plan and alert the Project Manager to such a need.

5.2 Project Manager

Prepare and submit to the Make-or-Buy Committee members lists of specific items or services requiring their Make-or-Buy decisions.

5.3 Make-or-Buy Committee

Review items submitted for decision and vote on them. The decision is by a majority vote.

5.4 Project Support Analyst

Insure that costs proposed to the customer are consistent with the Make-or-Buy Plan.

Publish and distribute Committee meeting minutes, including rational for decisions, to Committee members and any other individual designated by the Project Manager.

Maintain files of Make-or-Buy Committee actions.

5.5 Project Manager

Reconvene the Make-or-Buy Committee when circumstances necessitate a change in the Make-or-Buy Plan.

If the change is to a customer-approved Make-or-Buy Plan, requests the Contract Administrator to obtain customer approval before making the change.

5.6 Contract Administrator

Obtain customer approval of the change as requested.

5.7 Project Support Analyst

Initiate and maintain logs and records to document adherence to the Make-or-Buy Plan. Prepare any contractually required reports of the same.

6.0 MAJOR MAKE-OR-BUY ITEMS

6.1 Deliverable Equipment

The major items identified as "buy" items are those that have been designated by the Make-or-Buy Committee as major buy items. These items require procurement of proper, sufficient, product specifications or specification and source control drawings.

Title: **PRODUCT SAFETY**		Number: EP-1-4 Revision: A
Prepared by:	Approved by:	

1.0 PURPOSE

The purpose of this procedure is to define the method used to identify products that have a safety impact with respect to regulatory agencies.

2.0 APPLICABLE DOCUMENTS

None.

3.0 OVERVIEW

Consideration for the safety of operators and service personnel is of paramount importance in the design and manufacture of all products. All products are manufactured in accordance with the standards established by the Underwriters Laboratories, Canadian Standards Association and the Federal Communications Commission.

New products introduced after 1981, will be designed to comply with the standard of the International Electrotechnical Commissions (ICE) in those instances where there is no adverse economic impact.

Certain products are manufactured to ICE or Verband Deutscher Electrotechniker (VDE) standards when required by customers.

4.0 GENERAL

4.1 Integrity of Qualification by Product Safety Agencies

Company products are qualified by the product safety agencies when it is determined that they comply with design standards. Include only those components and materials that are authorized and have passed tests administered by those agencies.

Qualification applies only to the configuration submitted to the agencies for scrutiny. Configuration changes to features or components regulated by the product safety agencies can lead to withdrawal of qualification if the agencies have not first given their approval.

Product Safety agencies track components by the manufacture's name, model and catalog number for the component. Modification of a product safety affected component that results in assignment of a new name, model or catalog number by the vendor can lead to withdrawal of qualification if the agencies have not first given their approval to the use of the modified component.

4.2 Hazardous Material

Polychlorinated Biphenyl (PCB). Printed circuit board components or materials containing PCB will not be used in company products.

Cadmium will not be used in any product manufactured at the company.

4.3 Hazardous Material List

If any of the substances listed in the following paragraph are contained in components sold by the company, either as an integral part of the product or residue from the process, the vendor will so notify the company in writing so that proper entry can be made in the Safety Department Hazardous Material File.

Asbestos	Ethyleneimine
Coal tar pitch volatiles	2-Acetylaminofluorine
4-Nitrobiphenyl	4-Dimethylaminoazobenzene
Alpha-Naphthylamine	N-Nitrosodimethylamine
4,4'-Methylene bis (2-Chloroaniline)	Vinyl Chloride Monomer
Methyl Chloromethyl ether	Inorganic Arsenic
3,3' Dichlorobenzidine	Benzedine
bis-Chloromethyl ether	1,2-dibromo-3-Cloropropane (DBCP)
Beta-Naphthylamine	Acrylonitrile Monomer
4-Aminodiphenyl	Beryllium
	Beta-propiolactone

Compliance with the Toxic Substances Control Act is required for items containing any of the substances above.

Reference: OSHA Safety and Health Standard
29 CFR 1910.1001-1045

5.0 UNDERWRITERS LABORATORIES, INC.

Underwriters Laboratories (UL). Tests and evaluates products according to applicable UL Standards and then reports the results to the manufacture. UL tests products for foreseeable hazards to life and property. When a product meets UL Standards, it is entitled to carry the UL symbol.

5.1 UL Product Report

Documented proof that a product has met UL Standards if found in the UL Product Report published by UL.

This report consists of complete description of the product safety affected features of the product. All product safety affected components and materials are identified both in writing and in photographs that are part of the report. Test results and schematics of primary circuits are also included. A component or material may be added to the report by satisfying the requirements that pertain to that component or material. Submittals to UL are batched at 2-week intervals. A lead time of 8 weeks should be expected from the date of submittal to Safety to the receipt of approval or rejection by UL.

5.2 UL Listing

A factory produced product that has met UL Standards for end user application as a separate entity.

Manufactures's name and model numbers appear on a UL issued white card.

UL Listing Mark (Must be on the part.)

UL (This is not the official symbol.)

5.3 Traceability

Proof of UL Listing is appearance of UL Listing Mark or Underwriters Laboratories, Inc. (or abbreviation "UL") on the product. In some cases, UL allows marking by some other means. Vendors meeting UL Listing requirements will be found in the UL Electrical appliance and utilization Equipment Directory, or on a UL issued white listing card.

5.4 UL Recognition

A component that meets UL Standards for factory installation in a UL Listed or Recognized end product.

Manufacturer's name, model number, and marking requirements appear in the UL Recognized component directory.

Manufacturer's name, model number, and marking requirements appear on the issued yellow card.

UL Recognition Mark (Optional on the part.)

RU (This is not the offical symbol.)

5.5 Traceability

Only components which bear the recognized marking as specified in the UL Recognized Component Directory, or in the Recognized card file, should be considered to be Recognized Components. Proof of UL Recognition is appearance of the component manufacture's company name (or trade name symbol) and catalog designation on the component. In some cases UL allows marking by some other means; for instance, if the component is, in UL's judgment, too small to display the marking, the marking may appear on the box or carton containing components.

Note: *In cases where marking is not applied to the product, Receiving Inspection must record, in the part number folder, all marking and part information.*

5.6 UL Classification

A product that is evaluated only for specific hazards or under specific conditions. Classified components are neither listed nor recognized.

Example of UL Classification Mark on Classified component:

"Classified by Underwriters Laboratories, Inc. with respect to flammability. Class 2."

5.7 UL Follow-Up Services

A product that has met requirements in the UL Standard and has become Listed, Recognized, or Classified is subject to regular inspections by UL follow-up services. UL follow-up inspectors visit the company each quarter.

6.0 CANADIAN STANDARDS ASSOCIATION

Canadian Standards Association (CSA), like UL, tests and evaluates a product according to applicable CSA Standards.

The Safety Department conducts or supervises all tests and evaluations and prepares reports. A CSA Engineer visits the company each quarter to examine test and produce reports. CSA Certification marking can be affixed to new products as soon as the Safety Department completes the CSA Product report.

6.1 The CSA Product Report

Documented proof that a product has met CSA standards is found in the CSA Product Report. This report is similar to the UL product report described in Paragraph 5.1, except that it is prepared and published by the CSA Designated Representative of the Safety Department, and is reviewed and approved by CSA.

6.2 CSA Certification

All products that meet CSA Standards, regardless of usage, are certified. Component parts are CSA Certified with usage limitations. Manufacture's name, model numbers, marking requirements, and product information appear on a CSA issued white Certification card.

6.3 CSA Symbol (Must be on the part, except as noted below.)

CSA (This is not the offical symbol.)

6.4 Traceability

Proof of CSA Certification is the marking on the product as it is specified in the CSA List of Certified Electrical Equipment. In some cases, CSA allows marking by means other than on the product; for instance, if the product is, in CSAs judgement, too small to display the marking, the marking may appear on the box or carton containing the product. Marking requirements are on the Certification card.

Note: *In cases where the marking is not applied to the product, Receiving Inspection must record, in the part number folder, all marking and part information.*

7.0 VERBAND DEUTSCHER ELEKTROTECHNIKER

Verband Deutscher Elektrotechniker (VDE), like UL, tests and evaluates a product according to applicable VDE standards and reports the results to the manufacture. Production units that qualify for VDE approval must use VDE approved primary components. VDE approval of components can be obtained by having vendors submit VDE forms to the company so that they can accompany a product being submitted for approval. (See Paragraph 7.1.)

VDE Approval Mark (Not required on parts.)

VDE (This is not the offical symbol.)

7.1 VDE Forms

For motors and motor-operated appliances, VDE Form 43-E will be sent to the vendor to be filled out and signed.

For transformers, VDE Form 68-E will be sent to the vendor to be filled out and signed.

Completed VDE forms shall be sent to the Safety Department.

8.0 INTERNATIONAL ELECTROTECHNICAL COMMISSION

International Electrotechnical Commission (ICE) sets standards internationally. They do not conduct tests, but UL and VDE conduct tests according to ICE standards when requested by manufacturers wanting to achieve international acceptance of their products. Some VDE standards are identical to those of ICE. A world-wide effort is currently underway to harmonize UL, CSA, and other national standards with those of ICE.

8.1 ICE Classification

Products tested and passed by UL according to ICE standards may carry a UL Classification label indication compliance with the ICE standard.

Title: **ENGINEERING REVIEW BOARD**		Number: EP-1-5 Revision: A
Prepared by:	Approved by:	

1.0 PURPOSE

The purpose of this procedure is to define the method used to establish and use the Engineering Review Board to best utilize companies resources and whether to commit to enhancing old products or create new ones.

2.0 APPLICABLE DOCUMENTS

None.

3.0 OVERVIEW

The Engineering Review Board may be used for any of the following reasons:

1. Preliminary reviews of new product designs during the Development and Preprodction Phases.

2. Pre-release review of engineering documentation from Development to Preproduction or to Production phases.

3. Preliminary review of an Engineering Change Package.

4.0 PROCEDURE

The exact format for conducting the meeting will very depending on the objective of the documentation being reviewed, whether it be layout drawings, preliminary bills of materials, functional or design specifications, final drawings and bills of materials, etc., or new standards and procedures.

Several Engineering Review Board meetings may be required to define a new product prior to release for production.

The meeting shall be chaired by an engineering representative. Attendees will vary depending on the type of documentation to be reviewed.

4.1 Preliminary Documentation Review

(Informal) The following departments may be included on as needed bases: Engineering, Manufacturing, Purchasing and Quality.

The following are some suggested items that may be considered during the meeting:

1. Review new layout and preliminary documentation.
2. Review technical content of new designs.
3. Verify if correct approach.
4. Make decisions on how to proceed.

4.2 New Documentation Pre-release Review

(Formal) The following departments will be included:

Engineering, Manufacturing, Purchasing and Quality.

New Standards and Procedures. (Informal) Engineering review only. All other departments review and approve these documents in the Standard Review Committee meetings.

4.3 Changed Documentation Review

(Informal) This meeting is held prior to the Engineering Change preanalysis period. The primary elements to be considered at the meeting are:

1. The technical content of the change.

2. The verification of correct change documentation.

3. The engineering dispositions.

Title: **PRODUCT PHASES**		Number: EP-1-6 Revision: A
Prepared by:	Approved by:	

1.0 PURPOSE

The purpose of this procedure is to provide instructions, and to assign responsibilities for defining the various phases that all products go through during their life cycle, from development to production. It explains the kinds and amounts of documentation that are required for each phase.

2.0 APPLICABLE DOCUMENTS

Engineering Procedures:

EP-1-5, Engineering Review Board

EP-6-4, Engineering Change Procedure

EP-6-8, Change Control Board

EP-7-2, Document Release

3.0 OVERVIEW

There are three product phases that all products go through and they are Development, Preproduction and Production. The following paragraphs outline each of the phases.

4.0 DEFINITIONS

4.1 Form

The manufactured structure, shape, and material composition of an item or assembly.

4.2 Fit

The size and dimensional aspects of an item or assembly.

4.3 Function

The actual performance and level of performance of an item or assembly.

5.0 DEVELOPMENT PHASE

5.1 The Development phase is the first phase in the life cycle of a product. It continues until the product is developed to the point that it can be manufactured in small lots.

5.2 During the Development phase, sufficient feasibility must be demonstrated to justify further development effort. There should be some latitude in the design to avoid restrictions on both engineering and manufacturing, which may result in an unproducible or saleable product.

5.3 Documentation requirements during the Development phase shall state enough technical objectives for manufacturing to be able to produce the product with the aid of Engineering.

5.4 During the Development phase, Engineering has full control over product design and changes. The following items apply to change control during the Development phase:

1. The Engineering Change procedure is not required to make changes to development documentation. Revisions are made directly to the document. Change Control Board activity is not required.

2. Approval from other departments is not required to make changes.

3. If the change(s) affects form, fit, or function, the part number will change.

4. The revision level of the drawing will change each time there is a change to the part or drawing, beginning with P1, P2, etc.

5. A Document Release Notice form is required to release development documentation. (Reference: EP-7-2, Document Release.)

6.0 PREPRODUCTION

The Preproduction phase applies to products and parts which are under finalization of design by Engineering and which have been released to production in limited quantities. No more than 1000 items should be manufactured during the Preproduction phase. The following items apply to change control during the Preproduction phase.

1. More than one related problem may be fixed in each Engineering Change package, but must be listed separately on the Engineering Change Cover Sheet. (Reference: EP-6-4, Engineering Change Procedure.)

2. The Engineering Review Board may be required to review the Engineering Change package prior to sign off. (Reference: EP-1-5, Engineering Review Board.)

3. A Pre-analysis review of the Engineering Change package is required by all members of the Change Control Board. (Reference: EP-6-8, Change Control Board.)

4. The Change Control Board meeting and sign-off by Engineering is required prior to documentation release.

5. The part number is under Engineering Release status and only limited quantities may be purchased.

6. The part number must change if the Engineering Change affects the form, fit, or function of a product.

7. The revision level of a document will change with each Engineering Change, beginning with A for release to Engineering status, and subsequent changes to be B, C, D, etc.

8. All Preproduction Engineering Changes shall be stamped "Preproduction".

7.0 PRODUCTION PHASE

7.1 The Production phase shall be entered only after the product and it's design documentation have been verified as adequate for producing unlimited quantities of the product.

7.2 At the time of Production release to manufacturing, all engineering documentation supporting the product shall be put under change control. Copies of the drawings as they appeared before Production release shall be placed in a history file in Document Control.

7.3 The following items apply to change control during the Production phase:

1. More than one related problem may be covered in each Engineering Change, but must be listed separately on the Engineering Change cover sheet.

2. Review of Engineering Changes are required by members of the Change Control Board.

3. Change Control Board meeting and signing of Engineering Changes by all members is required prior to documentation release to Manufacturing.

4. If the change affects form, fit, or function of the part, the part number shall change.

5. The revision level of the drawing will change each time there is a change to the part or drawing, beginning with A, B, C, etc.

6. An Engineering Change form and a Document Release Notice form are required for Production released documentation.

Title: **DESIGN REVIEWS**		Number: EP-1-7 Revision: A
Prepared by:	Approved by:	

1.0 PURPOSE

The purpose of this procedure is to explain how to establish and use design reviews to provide assurance that products are designed so that they can be manufactured and maintained through their intended life in a cost-effective manner.

2.0 APPLICABLE DOCUMENTS

Engineering Procedures:

EP-4-3, Deviation/Waiver

3.0 DEFINITIONS

3.1 Design Reviews

A series of reviews to assure that reliability considerations are incorporated in the proposed designs for new product or major revision to an existing product. These reviews are conducted in parallel with the design effort to minimize schedule impacts. This process includes the following specific reviews:

3.1.1 Preliminary Design Review. Review held shortly after the design concepts are completed to critique the conceptual design.

Title: **DESIGN REVIEWS**	Number: EP-1-7

3.1.2 Intermediate Design Review. Review held during the Development phase to review design layouts, preliminary drawings, and component selections (after breadboards have been made and tested).

3.1.3 Final Design Review. Review held during Preproduction phase and during which all drawings and specifications are reviewed. All open action items must be completed before final production release. This review usually includes results of prototype testing and preproduction testing.

4.0 OVERVIEW

The Program Manager has final responsibility for determining if design problems exist and for deciding upon the appropriate course of action to take.

5.0 RESPONSIBILITIES

5.1 Engineering is responsible for appointing a Program Manager for each new product under development and each major change to an existing product.

5.2 Program Manager

1. Schedule Design Review meetings at appropriate times during the development program and inviting representatives from the following areas, as appropriate:

   Engineering
   Manufacturing
   Purchasing
   Quality Assurance
   Other product design groups, as required

2. Chairing the Design Review meetings and assigning action items to appropriate individuals.

3. Assuring that action items identified at the Design Review meetings are recorded and communicated to all meeting attendees, division vice presidents, the Executive Vice President of Operations, and to all individuals responsible for action items.

4. Developing and enforcing an action plan for each action item.

5. Ensuring personally that no major action items remain unresolved after the first customer shipment. Deviations must be submitted in writing for approval by Quality and Field Engineering. (Reference: EP-4-3, Deviation/Waiver.)

5.3 Other Design Review participants (as defined in Item 1.).

5.3.1 Ensuring that every design item relating to their field of expertise is thoroughly reviewed.

5.5 Manufacturing Manager

5.5.1 Incorporating reports of Design Reviews into the appropriate Phase Reviews.

Title: **BASELINE MANAGEMENT**		Number: EP-1-8 Revision: A
Prepared by:	Approved by:	

1.0 PURPOSE

The purpose of this procedure is to define the techniques used to control production via specifications and drawings when multiple distinctively different versions of a product are to be manufactured.

2.0 APPLICABLE DOCUMENTS

Engineering Procedure:

EP-2-2, Document Number Assignment Logbook

Engineering Form:

E002, Configuration Baseline Document

E003, As-Built Configuration Record

3.0 GENERAL REQUIREMENTS

Documented baselines will be used as a common reference for change control of technical requirements and product configuration. The configuration baseline document (CBD) and as-built configuration record (ABCR) shall define, respectively, the as-designed and as-built product configuration baselines.

4.0 CONFIGURATION CONTROL PROCEDURE

Control of the configuration of an end item shall be established by means of incremental and summary reviews and progressive baseline development. As a product of the reviews, the CBD for each End Item will be prepare and maintain and, when applicable, its series of configurations.

This procedure is divided into the following parts:

- Configuration Baseline Document
- Configuration Baseline Document Form Preparation
- As-Built Configuration Record
- As-Built Configuration Record Form Preparation

4.1 Configuration Baseline Document

The CBD shall consist of a listing of all significant hardware/software configuration documents (specification, drawings, and lists) that define an End Item. A CBD will be prepared for each deliverable End Item and when changes are authorized, include subsequent configurations of the basic requirement. (See Figure 2. for an example of the form.)

4.1.1 CBD Control. The CBD (including revisions) will be forwarded to the customer, if required. All changes to the imposed baseline shall be subject to formal change control and will be processed as Class 1 changes. The CBD and all authorized outstanding changes will establish the current baseline at any point in time.

4.1.2 CBD Maintenance. The CBD will be maintained in accordance with standard procedures. The CBD will be prepared and forwarded to the customer, if required, for approval of the progressive baseline documentation definition as an output of design reviews, configuration inspection, or product audit.

4.1.3 CBD Content. The following specific elements of data shall be provided:

1. Identification of the End Item or major component by a customer control number or part number as applicable, item nomenclature, and part number.

2. The customer's procurement technical documentation (drawing or specification) or procurement documentation, Acceptance Test Procedure or specification, by title or number.

3. The released baseline drawing list, by assembly indenture, for the item.

4. Identify the appropriate traceability code (i.e., T_s - Traceability by serial number, E - Exempt from traceability) for the nameplated items or assemblies as directed by the customer's procurement document or design document.

5. Selected procurement documentation and Materials and Processes Specifications that are to be baselined for the End Item or major components, will be delivered to the customer. These documents may be listed integral with or separately from the assembly indenture data, including the change letter, release date and change authority.

CONFIGURATION BASELINE DOCUMENT						
Engineer: ❶			Date:		Configuration Baseline Document No: ❷ Revision:	
Customer: ❸			Date:		Date:	
Indent No.	Document No.	Description	Qty	Traceability Code	Rev.	Eng Change No.
❹	❺	❻	❼	❽	❾	❿

Form E002 (Procedure EP-1-8)

Figure 2. Configuration Baseline Document Form

4.1.4 Configuration Baseline Document Form Preparation

The procedure for processing the Configuration Baseline Document form shall be followed by each individual responsible for entering information on the form. Each circled number below corresponds to the circled number on the Configuration Baseline Document form E002.

4.1.5 Originator

❶ Enter your name and the Date that the form was completed.

4.1.6 Document Control

❷ Assign the next available number from the Document Number Assignment Logbook EP-2-2, under Paragraph 13.0, Configuration Baseline Document Number Assignment Log. Then enter the number, Revision Letter, and Date in the Configuration Baseline Document form E002.

4.1.7 Customer

❸ Enter your name, Company Name and the Date.

4.1.5 Originator

❹ Enter part indentured number.

❺ Enter Document number.

❻ Enter Part description.

❼ Enter Quantity per assembly.

❽ Enter Serialized assemblies indicated by identification and traceability code.

❾ Enter the revision letter of part.

❿ Enter all Engineering Change numbers.

4.2 As-Built Configuration Record

The ABCR for each End Item or major component, are the documented and verified identification and physical description of each End Item or major component manufactured. The ABCR consists of an accurate accumulation of data extracted from the fabrication and test records, as certified by inspection records. Listings of standard hardware (i.e., nuts, bolts, washers, etc.), is not required unless these parts are either 1) critical, 2) special application hardware or 3) require traceability. (See Figure 3. for an example of the form.)

4.2.1 As-Built Configuration Record Content

The following specific elements of data will be forwarded to the customer, if required, for each End Item and its subordinate parts.

1. Identification of the End Item or major component by the customer control number or part number, item nomenclature, and part number, and serial number.

2. The procurement technical documentation, Acceptance Test Procedure or specification, by title and number.

3. The ABCR record consisting of an accurate indentured parts list of the completed End Item as fabricated and tested by Manufacturing and certified by inspection records. Changes incorporated shall be identified by the appropriate engineering change orders. This record shall also include the record of authorized deviations or waivers by part number.

AS-BUILT CONFIGURATION RECORD						
Engineer: ❶			Date:	As-Built Configuration Record No: ❷		
Manufacturing: ❸			Date:	Revision: Date:		
Indent No.	Document No.	Description	Qty	Serial No.	EC No.	As-Designed/ As-Built Differences
❹	❺	❻	❼	❽	❾	❿

Form E003 (Procedure EP-1-8)

Figure 3. As-Built Configuration Record Form

4.2.2 As-Built Configuration Record Form Preparation

The procedure for processing the As-Built Configuration Record form shall be followed by each individual responsible for entering information on the form. Each circled number below corresponds to the circled number on the As-Built Configuration Record form E003.

4.2.3 Originator

❶ Enter your name and the Date that the form was completed.

4.1.6 Document Control

❷ Assign the next available number from the Document Number Assignment Logbook EP-2-2, under Paragraph 14.0, As-Built Configuration Record Number Assignment Log. Then enter the number, Revision Letter, and Date in the As-Built Configuration Record form E003.

4.1.7 Manufacturing

❸ Enter your name and the Date the form was completed.

4.1.5 Originator

❹ Enter part indentured number.

❺ Enter Document number.

❻ Enter Part description.

❼ Enter Quantity per assembly.

❽ Enter Serial Number.

❾ Enter Engineering Change Number(s).

❿ Enter the open and unincorporated engineering changes that are outstanding for the applicable part or assembly number.

Title: **PRODUCT STRUCTURING**		Number: EP-1-9 Revision: A
Prepared by:	Approved by:	

1.0 PURPOSE

The purpose of this procedure is to standardize the product structuring methods to be used. It will define the procedures to be used when structuring by either the Feature Code or Top Level Assembly methods.

2.0 APPLICABLE DOCUMENTS

None.

3.0 PRODUCT STRUCTURE

A product structure refers to all bills of materials required to produce a complete end product. It also refers to how the bills of materials are related. Bills of materials are like building blocks. Detail parts are systematically fastened together to form subassemblies. These subassemblies are combined with other assemblies to create units and end products. Each assembly must have its own part number. When structuring a product, one of the methods listed in Paragraph 3.1 or 3.2 shall be used.

3.1 Feature Code Method

The Feature Code method of structuring will be used when it is known through product planning that the end item will be a family of products which will have a basic configuration with several variable features.

The Feature Code System begins with the definition of a basic end product which contains all of the parts that remain constant regardless of the ultimate configuration. The remaining parts are grouped under feature code Bills of Materials according to their function.

3.1.1 Selective Feature. A selective feature is a product function for which there may be several alternatives; however, each product must have one of the available choices in order to be operational. In the example below, it has been determined that the engine will be a selective feature. (See Figure 4.)

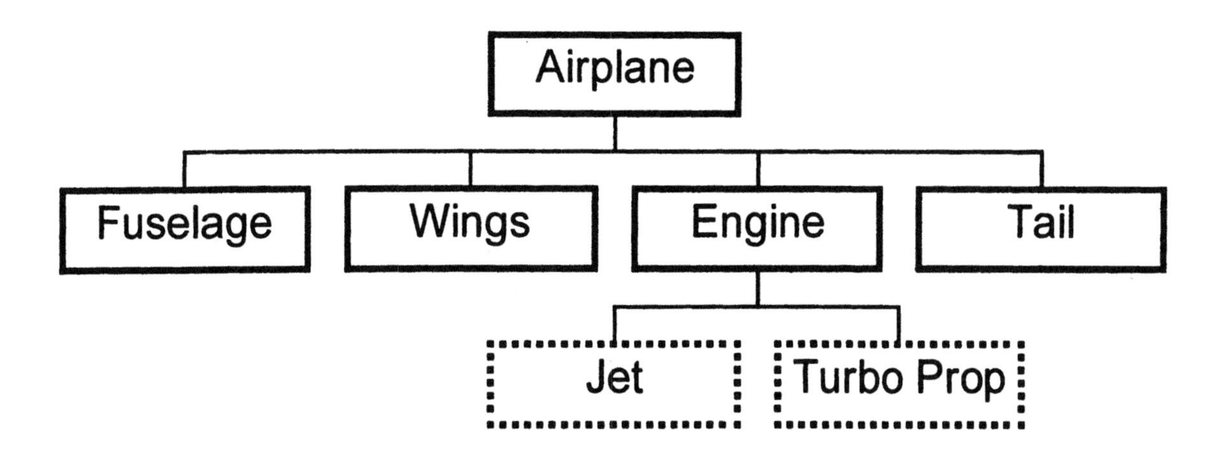

Figure 4. Selective Feature

3.1.2 Optional Feature. An optional feature is a product function beyond the basic operation configuration. It is generally available to the customer at an additional charge. In the example below it has been determined that the spare tire will be an optional feature. (See Figure 5.)

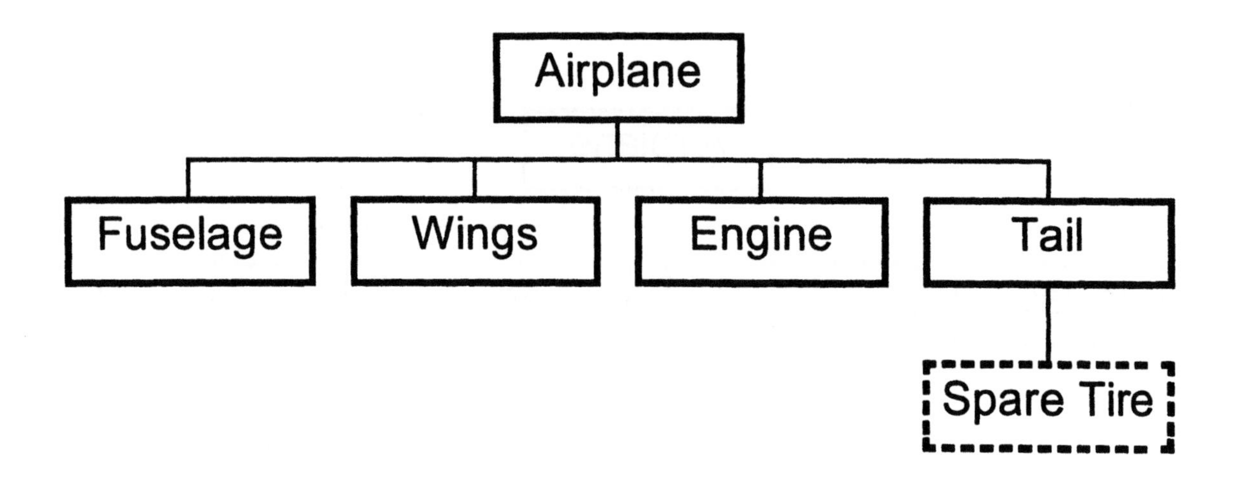

Figure 5. Optional Feature

3.2 Top Level Assembly Method

The top level assembly method will be used when it is determined through Product Planning that there will be few, if any, variations from the standard. A Top Level Assembly is a conventional structure of parts which contains all the parts used in the end item. There are no variations definable within that Top Level Assembly. If a similar but different configuration is required, it must be defined by its own Top Level Assembly. The following example depicts a Top Level Assembly structure.

The Top Level Assembly System is a method of structuring a bill of materials where all component parts and assemblies are structured in such a manner as to lead to one single Top Level Assembly number.

3.2.1 Drawing Tree. A drawing tree will be generated which will depict the relationship of the assemblies used. This will be a pictorial representative of a complete structure of the end product. (See Figure 6.) The example below represents a complete end product assembly.

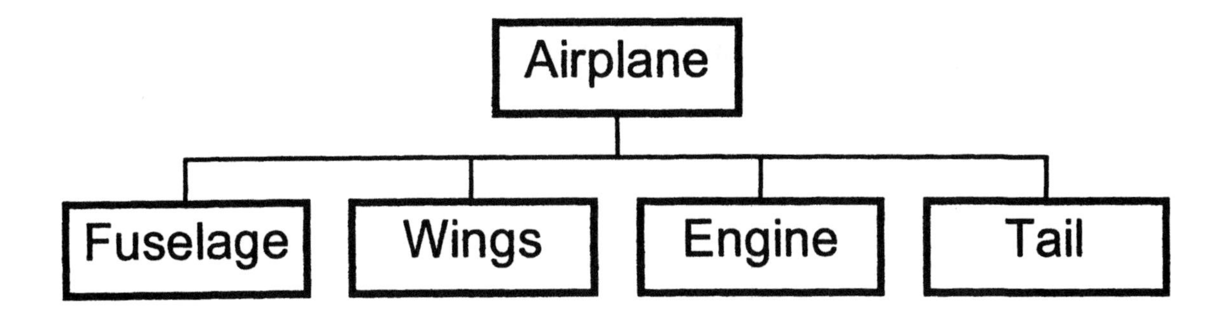

Figure 6. Top Level Assembly

Title: **COMPATIBILITY BOOKS**		Number: EP-1-10 Revision: A
Prepared by:	Approved by:	

1.0 PURPOSE

The purpose of this procedure is to explain how to develop compatibility books that will retain information that applies to the compatibility of assemblies, where incompatible conditions may exist due to engineering change activity or model differences.

2.0 APPLICABLE DOCUMENTS

None.

3.0 DEFINITIONS

3.1 Downward Compatible

A new assembly (usually a higher part number) is a direct replacement (downward compatible) for an old assembly (usually a lower part number) when it does not degrade the product performance. Downward compatibility is indicated when both assemblies have the same compatibility number.

3.2 Upward Compatible

A old assembly (usually a lower part number) is equivalent to (upward compatible) and may directly replace a new assembly (usually a higher part number) if its use does not affect product performance. Upward compatibility is indicated when assemblies having the same compatibility number are also indicated as equivalent in the description of the Engineering Change.

4.0 OVERVIEW

4.1 Compatibility books list field replaceable units that are determined by Engineering, such as printed circuit cards, power supplies, micro code, etc., and summarizes the problems that were fixed. They also identify the evolution from one design to another.

5.0 COMPATIBILITY BOOKS PROCEDURE

5.1 For example, each page in a compatibility book contains the compatibility information for one printed circuit board type. It shows the history of the engineering changes that resulted in part number changes for that printed circuit board. It also shows which part numbers are obsolete and which part numbers can be used to replace them.

5.2 For some printed circuit boards, the functional changes that resulted in new part numbers will not function without corresponding product rework. This usually results in a compatibility number change for the new assembly. The description of the engineering change will summarize the changes. The Field Bill of Material defining the required rework will be listed in a separate column next to the engineering change number.

5.3 If the compatibility number does not change, the new printed circuit board will perform at least the same functions as the older printed circuit board.

6.0 COMPATIBILITY AND OBSOLESCENCE

6.1 Any part number that is prefixed with an asterisk is an obsolete part number. Obsolete means that the part will no longer be supplied. If an obsolete part is ordered, the Field Parts Crib will substitute a part that is downward compatible or equivalent to the part number ordered.

6.2 For some printed circuit boards, all part numbers of the printed circuit board at a particular compatibility number have been obsoleted, leaving no available replacements at that compatibility number. This situation occurs when a printed circuit board is obsoleted by a mandatory engineering change.

6.2.1 If replacement is ordered for an obsolete printed circuit board for which there is no replacement at the same compatibility number, it is possible that one or more mandatory engineering changes are missing from the product. Check the engineering change log for that product and the compatibility page for that printed circuit board for a description of the engineering change and the applicable bill of material numbers.

6.2.2 Update the product so that a replacement printed circuit board of a higher compatibility number can be used. (It may be necessary to order the indicated assembly parts list which will produce the correct printed circuit board that is needed.)

6.3 Downward compatibility is designated by the compatibility number to the right of the part number. A printed circuit board of a given part number is downward compatible with all of those lower part number printed circuit boards that have the same compatibility number. This means it can be used to replace any of the lower part number printed circuit boards having the same compatibility number without requiring rework to the printed circuit board or the product.

Title: **COMPATIBILITY BOOKS**	Number: EP-1-10

For Example:

PCB Number	Description	Engineering Change No.
*42150 100	Initial Release - W/W Board	40020
*42150 101	Corrects Assembly Parts List - Equivalent to 42150 100	40030
42150 210	Release of Printed Circuit Board Layout - Added new logic to Data check circuitry. Equivalent to 42150 101.	8612
42150 220	Corrects timing problem. Contains wiring to implement logic changes that created this part number.	8619
42154 300	Release of Printed Circuit Board Layout.	None.

In the above example:

Part number 42150 100 can be replaced by itself, part number 42150 101 or 42150 210.

Part number 42150 101 can be replaced by itself, part number 42150 100 or 42150 210.

Part number 42150 210 can be replaced by itself, part number 42150 100 or 42150 101.

Part number 42150 220 can replace itself and all part numbers less than itself as long as the family number stays the same.

Part number 42154 300 can replace only itself.

In addition, since part number 42150 100 and part number 42150 101 have an * to the left of the part number they have been obsoleted. However, the Field Crib can replace them with any part number which has the same compatibility number.

Title: **DRAWING TREES**		Number: EP-1-11 Revision: A
Prepared by:	Approved by:	

1.0 PURPOSE

The purpose of this procedure is to show an example of a drawing tree that is used for visibility and planning purposes.

2.0 APPLICABLE DOCUMENTS

None.

3.0 DRAWING TREE

A drawing tree depicts the relationship of the assemblies and subassemblies used. This will be a pictorial representative of a complete structure of the end product. (See Figure 7.) The example below represents just part of an end product.

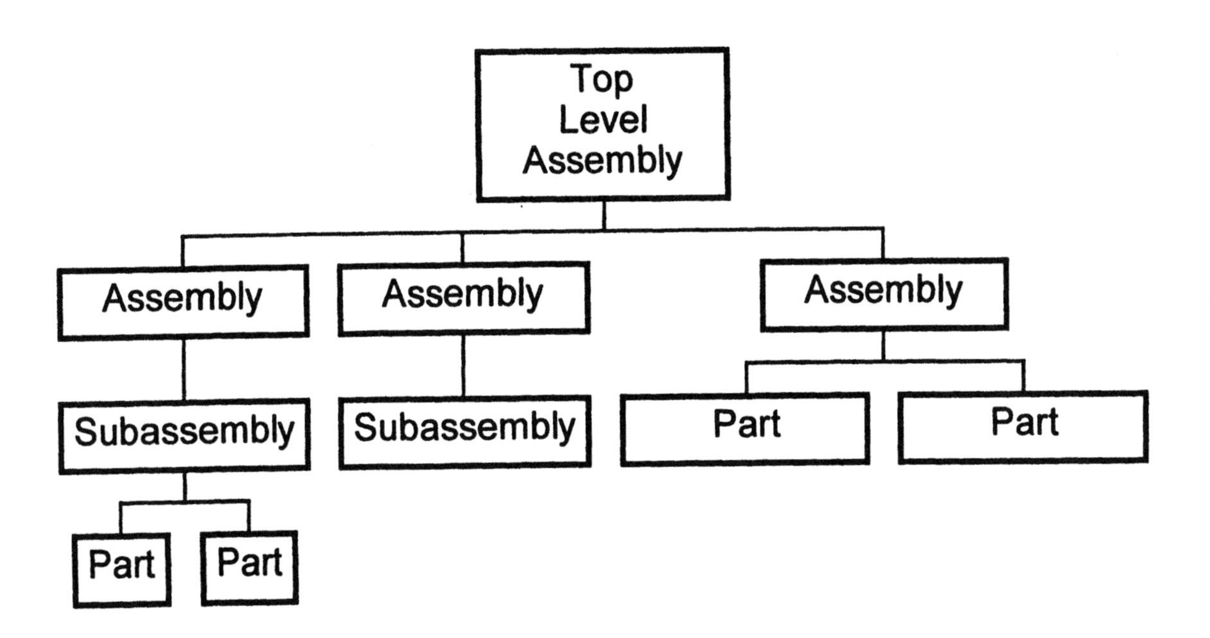

Figure 7. Drawing Tree

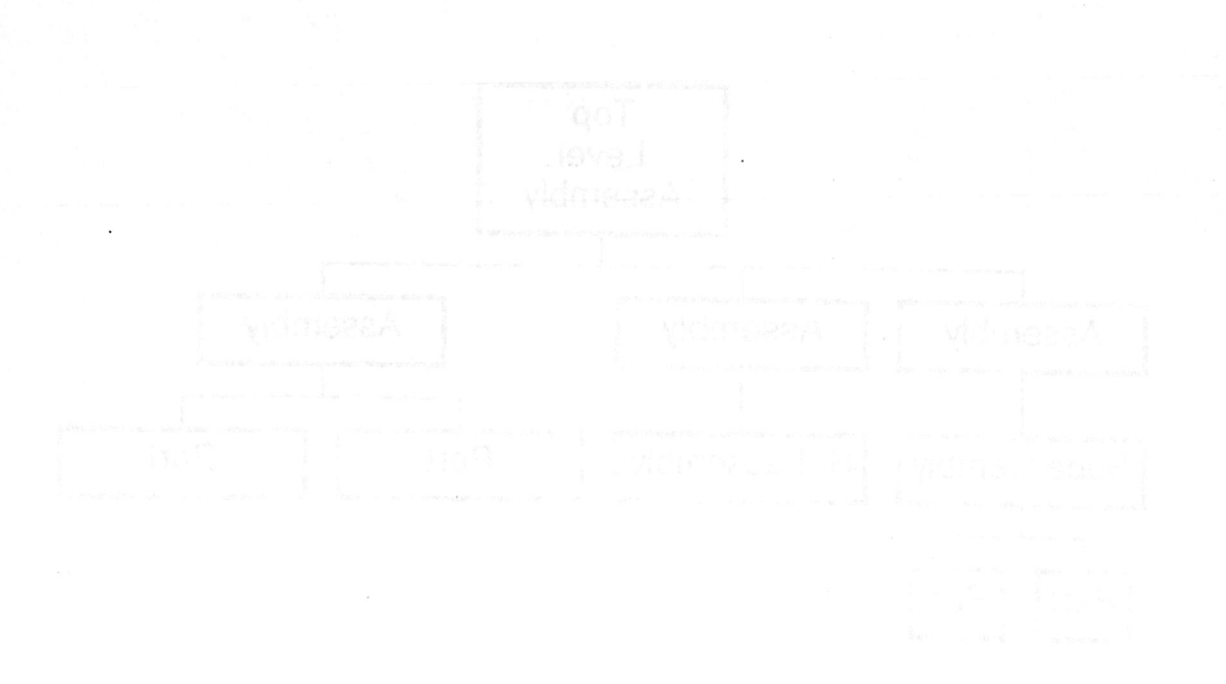

SECTION 2

PRODUCT AND DOCUMENTATION IDENTIFICATION

Title: **DOCUMENT NUMBERING SYSTEM**		Number: EP-2-1 Revision: A
Prepared by:	Approved by:	

1.0 PURPOSE

The purpose of this document is to establish the method used to assign identifying numbers to parts, assemblies, and related engineering documentation.

2.0 APPLICABLE DOCUMENTS

Engineering Procedures:

EP-2-2, Document Number Assignment Logbook

EP-2-3, Block Number Assignment

3.0 DEFINITIONS

3.1 Part

A part is any single inseparable item, or separable unit or assembly, whose design parameters are defined by documentation prepared by the company and controlled by Document Control. Examples of parts are items such as:

Components (fabricated or procured)
Assemblies
Media (diskettes, disks, tapes, etc.)
Bulk items (paint, cable, etc.)

3.2 Document

A document is any form of media which is prepared and released for the purpose of defining and controlling the manufacture of company products. Following are the documents that fall in this category:

Plans
Reports
Drawings
Bills of Materials
Product Specifications
Process Procedure
Assembly Procedures
Material Procedures
Test Procedures
Cleaning Procedures
Material Handling Procedures
Packaging Procedures
Shipping Procedures

3.3 Document Type Designation

A two letter code used to identify the above listed documentation. Examples are: BM for bill of material, RP for report, etc.

4.0 DOCUMENT NUMBERING SYSTEMS

Following are examples of the document numbering method for each of the above listed documents. New document numbers are signed out from a logbook that resides in Document Control. (Ref. EP-2-2 Document Number Assignment Logbook.)

Note: *The spaces between the letters and numbers in the following examples are used to show the different elements of each number. The actual number does not have spaces.*

4.1 Product Specification

PS 123 (PD123 actual number)

- three digit sequential logbook number
- two letter document type designation

4.2 Assembly Drawings

12345 -00 (12345-00 actual number)

- two digit dash number
- five digit sequential logbook number

4.3 Detail Drawings

12345 -00 (12345-00 actual number)

- two digit dash number
- five digit sequential logbook number

4.4 Sketches

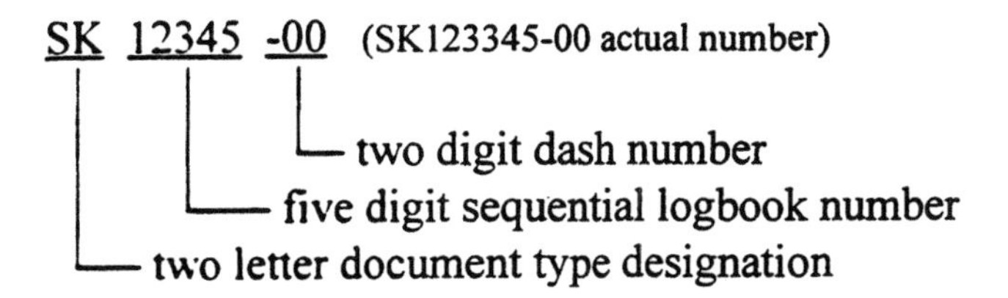

4.5 CAD Drawings

The CAD drawing number is used for magnetic (computer) storage and retrieval only. This number does not appear on the drawing.

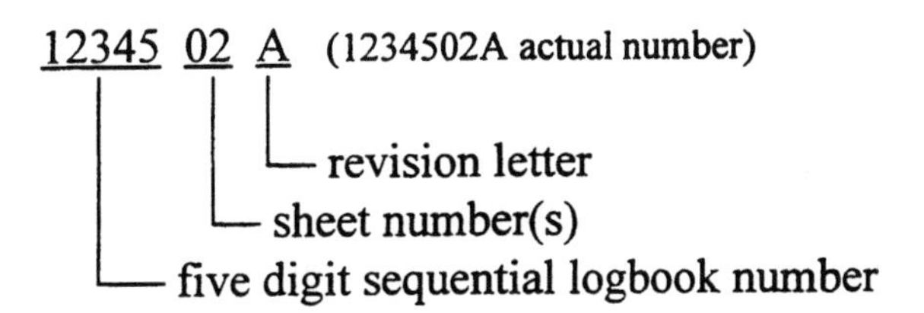

4.6 Facility Drawings

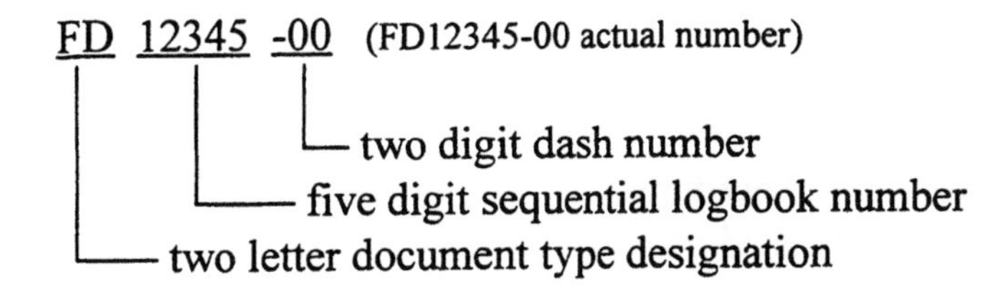

4.7 Bill of Material

A bill of material requires an identification number when it is prepared separate from the drawing.

```
BM  12345  -00   (BM12345-00 actual number)
|   |      |
|   |      └─ two digit dash number*
|   └─ five digit sequential logbook number*
└─ two letter document type designation
```

* This number is the same as the assembly drawing number.

4.8 Plans, Reports, Specifications and Standards

```
PL  12345   (PL12345 actual number)
|   |
|   └─ five digit sequential logbook number
└─ two letter document type designation (See examples.)
```

Examples:

PL . . . Plan
RP . . . Report
PS . . . Product Specification
PP . . . Process Procedure
AP . . . Assembly Procedure
MP . . . Material Procedure
TP . . . Test Procedure
CP . . . Cleaning Procedure
MH . . . Material Handling Procedure
PA . . . Packaging Procedure
SP . . . Shipping Procedure

4.9 Engineering Lab Notebooks

EL 001 (EL001 actual number)

- 001 — three digit sequential logbook number
- EL — two letter document type designation

4.10 Manuals

PM 001 (PM001 actual number)

- 001 — three digit sequential logbook number
- PM — two, three or four letter document type designation

Examples:

PM. . . Policy Manual
OPM. . Office Procedures Manual
DSM. . Documentation Standard Manual
AM. . . Accounting Manual
HRM. . Human Resources Manual
SM. . . Sales Manual
MK. . . Marketing Manual
PU. . . Purchasing Manual
QM. . . Quality Manual
QPM . . Quality Procedures Manual
SA. . . Safety Manual
MM . . Manufacturing Manual
MPM . Manufacturing Procedures Manual
EPM . . Engineering Procedures Manual
DSM . . Drafting Standard Manual

4.11 Manual Subsections

```
PM -1 -1 (PM-1-1 actual number)
|   |  |
|   |  └── one digit document number
|   └── one digit manual section number
└── two letter department designation (See examples.)
```

Examples:

PM . . . Policy Manual
OP . . . Office Procedures Manual
DSM . . . Documentation Standard Manual
AM . . . Administrative Manuals Control
FC . . . Forms Control
AG . . . Abbreviations/Glossary
AM . . . Accounting Manual
HRM . . . Human Resources Manual
SM . . . Sales Manual
MK . . . Marketing Manual
PM . . . Purchasing Manual
QM . . . Quality Manual
QPM . . . Quality Procedures Manual
QS . . . Quality Standard Practices
QP . . . Quality Special Practices
TS . . . Test Standard Practices
SA . . . Safety Manual
MM . . . Manufacturing Manual
MPM . . Mfg Procedures Manual
WS . . . Workmanship
EP . . Engineering Procedures Manual
DSM . . . Drafting Standard Manual

4.12 Forms

E 001 (E001 actual number)

- 001 — three digit sequential logbook number
- E — one letter department designation (See examples.)

Examples:

E . . . Engineering
M . . . Manufacturing
P . . . Purchasing
Q . . . Quality
S . . . Sales

Title: **DOCUMENT NUMBER ASSIGNMENT LOGBOOK**		Number: EP-2-2 Revision: A
Prepared by:	Approved by:	

1.0 PURPOSE

The purpose of this procedure is to provide a central location for assigning document numbers for product documentation.

1.1 APPLICABLE DOCUMENTS

Engineering Procedures:

EP-2-1, Document Numbering System

EP-2-3, Block Number Assignment

1.2 OVERVIEW

Document numbers are signed out of a logbook that resides in Document Control. A Document Master List of all numbers that have been assigned out of the Document Number Assignment Logbook will be distributed to all end users.

2.0 ENGINEERING LAB NOTEBOOK NUMBER ASSIGNMENT LOG

2.1 Engineering Lab Notebook Number Assignment Log Form Preparation

The procedure for processing the Engineering Lab Notebook Number Assignment Log form shall be followed by each individual responsible for entering information on the form. Each circled number below corresponds to the circled number on the Engineering Lab Notebook Number Assignment Log form E004. (See Figure 8. for an example of the form.)

2.1.1 Document Control

❶ Enter Originators name obtained form the Engineering Lab Notebook Request form E027. (Reference: EP-3-4, Engineering Lab Notebook.)

❷ Enter the date that the Engineering Lab Notebook was checked out from Document Control.

❸ Enter date the Engineering Lab Notebook was returned to Document Control.

❹ After the Engineering Lab Notebook is complete, enter the time span of the notebook from the start date to the last date an entry was made.

ENGINEERING LAB NOTEBOOK NUMBER ASSIGNMENT LOG				
Eng. Lab Notebook	Name	Date Checked out	Date Returned	Time Span From To
001	❶	❷	❸	❹
002				
003				
004				
005				
006				
007				
008				
009				
010				
011				
012				
013				
014				
015				
016				
017				
018				
019				
020				
021				
022				
023				
024				
025				
026				
027				

Form E004 (Procedure EP-2-2)

Figure 8. Engineering Lab Notebook Number Assignment Log Form

3.0 DOCUMENT RELEASE NOTICE NUMBER ASSIGNMENT LOG

3.1 Document Release Notice Number Assignment Log Form Preparation

The procedure for processing the Document Release Notice Number Assignment Log form shall be followed by each individual responsible for entering information on the form. Each circled number below corresponds to the circled number on the Document Release Notice Number Assignment Log form E005. (See Figure 9. for an example of the form.)

3.1.1 Document Control

❶ Enter the document number and title for all documents that are being released.

❷ Enter the associated Document Release Notice Number that was assigned from EP-2-2, Document Release.

❸ Enter your Initials.

❹ Enter the date that the Document Release number was assigned.

DOCUMENT RELEASE NOTICE NUMBER ASSIGNMENT LOG				
Document Release Notice No.	❶ Document Number and Title	❷ Engineering Change No.	❸ Initials	❹ Date
001				
002				
003				
004				
005				
006				
007				
008				
009				
010				
011				
012				
013				
014				
015				
016				
017				
018				
019				
020				
021				
022				
023				
024				
025				

Form E005 (Procedure EP-2-2)

Figure 9. Document Release Number Assignment Log Form

4.0 PRODUCT INTRODUCTION NUMBER ASSIGNMENT LOG

4.1 Product Introduction Number Assignment Log Form Preparation

The procedure for processing the Product Introduction Number Assignment Log form shall be followed by each individual responsible for entering information on the form. Each circled number below corresponds to the circled number on the Product Introduction Number Assignment Log form E006. (See Figure 10. for an example of the form.)

4.1.1 Document Control

Obtain the following information from the Product Introduction Notice form E001. (Reference: EP-1-2, Product Introduction.)

❶ Enter the Project Number. This number is developed using the method defined in EP-1-2, Product Introduction.

❷ Enter the project title.

❸ Enter the name of the product.

❹ Enter your initials.

❺ Enter the date that the Product Introduction Number was assigned.

PRODUCT INTRODUCTION NUMBER ASSIGNMENT LOG				
❶ Project No.	❷ Project Title	❸ Product	❹ Initials	❺ Date

Form E006 (Procedure EP-2-2)

Figure 10. Product Introduction Number Assignment Log Form

5.0 PART NUMBER ASSIGNMENT LOG

5.1 Part Number Assignment Log Form Preparation

The procedure for processing the Part Number Assignment Log form shall be followed by each individual responsible for entering information on the form. Each circled number below corresponds to the circled number on the Product Number Assignment Log form E007. (See Figure 11. for an example of the form.)

5.1.1 Document Control

Obtain the following information from the Part Number Request form E023. (Reference: EP-2-5, Part Numbering System.)

❶ Enter the part description.

❷ Enter your initials.

❸ Enter the date that the part number was assigned.

PART NUMBER ASSIGNMENT LOG			
Part Number	❶ Description	❷ Initials	❸ Date
00001-00			
00002-00			
00003-00			
00004-00			
00005-00			
00006-00			
00007-00			
00008-00			
00009-00			
00010-00			
00011-00			
00012-00			
00013-00			
00014-00			
00015-00			
00016-00			
00017-00			
00018-00			
00019-00			
00020-00			
00021-00			
00022-00			
00023-00			
00024-00			
00025-00			
00026-00			
00027-00			

Form E007 (Procedure EP-2-2)

Figure 11. Part Number Assignment Log Form

6.0 DEVIATION/WAIVER NUMBER ASSIGNMENT LOG

6.1 Deviation/Waiver Number Assignment Log Form Preparation

The procedure for processing the Deviation/Waiver Number Assignment Log form shall be followed by each individual responsible for entering information on the form. Each circled number below corresponds to the circled number on the Deviation/Waiver Number Assignment Log form E008. (See Figure 12. for an example of the form.)

6.1.1 Document Control

Obtain the following information from the Deviation/Waiver Request form E032. (Reference: EP-4-3, Deviation/Waiver.)

❶ Enter the part number and description.

❷ Enter your initials.

❸ Enter the date that the Deviation/Waiver number was assigned.

DEVIATION/WAIVER NUMBER ASSIGNMENT LOG			
Deviation/ Waiver No.	❶ Part Number and Description	❷ Initials	❸ Date
001			
002			
003			
004			
005			
006			
007			
008			
009			
010			
011			
012			
013			
014			
015			
016			
017			
018			
019			
020			
021			
022			
023			
024			
025			
026			
027			

Form E008 (Procedure EP-2-2)

Figure 12. Deviation/Waiver Number Assignment Log Form

7.0 LIMITED BUY AUTHORIZATION NUMBER ASSIGNMENT LOG

7.1 Limited Buy Authorization Assignment Log Form Preparation

The procedure for processing the Limited Buy Authorization Number Assignment Log form shall be followed by each individual responsible for entering information on the form. Each circled number below corresponds to the circled number on the Limited Buy Authorization Number Assignment Log form E009. (See Figure 13. for an example of the form.)

7.1.1 Document Control

Obtain the following information from the Limited Buy Authorization form E034. (Reference: EP-5-3, Limited Buy Authorization.)

❶ Enter the part number and description.

❷ Enter your initials.

❸ Enter the date that the Limited Buy Authorization number was assigned.

LIMITED BUY AUTHORIZATION NUMBER ASSIGNMENT LOG			
Limited Buy Authorization No.	❶ Part Number and Description	❷ Initials	❸ Date
001			
002			
003			
004			
005			
006			
007			
008			
009			
010			
011			
012			
013			
014			
015			
016			
017			
018			
019			
020			
021			
022			
023			
024			
025			
026			
027			

Form E009 (Procedure EP-2-2)

Figure 13. Limited Buy Authorization Number Assignment Log Form

8.0 SINGLE SOURCE AUTHORIZATION NUMBER ASSIGNMENT LOG

8.1 Single Source Authorization Assignment Log Form Preparation

The procedure for processing the Single Authorization Number Assignment Log form shall be followed by each individual responsible for entering information on the form. Each circled number below corresponds to the circled number on the Single Source Authorization Number Assignment Log form E010. (See Figure 14. for an example of the form.)

8.1.1 Document Control

Obtain the following information from the Single Source Authorization form E033. (Reference: EP-5-2, Single Source Authorization.)

❶ Enter the part number and description.

❷ Enter your initials.

❸ Enter the date that the Single Source Authorization number was assigned.

SINGLE SOURCE AUTHORIZATION NUMBER ASSIGNMENT LOG			
Single Source Authorization No.	❶ Part Number and Description	❷ Initials	❸ Date
001			
002			
003			
004			
005			
006			
007			
008			
009			
010			
011			
012			
013			
014			
015			
016			
017			
018			
019			
020			
021			
022			
023			
024			
025			
026			
027			

Form E010 (Procedure EP-2-2)

Figure 14. Single Source Authorization Number Assignment Log Form

9.0 ENGINEERING CHANGE PROPOSAL NUMBER ASSIGNMENT LOG

9.1 Engineering Change Proposal Number Assignment Log Form Preparation

The procedure for processing the Engineering Change Proposal Number Assignment Log form shall be followed by each individual responsible for entering information on the form. Each circled number below corresponds to the circled number on the Engineering Change Proposal Number Assignment Log form E011. (See Figure 15. for an example of the form.)

9.1.1 Document Control

Obtain the following information from the Engineering Change Proposal form E031. (Reference: EP-4-2, Engineering Change Proposal.)

❶ Enter the part number and description.

❷ Enter your initials.

❸ Enter the date that the Engineering Change Proposal number was assigned.

ENGINEERING CHANGE PROPOSAL NUMBER ASSIGNMENT LOG			
Engineering Change Proposal No.	❶ Part Number and Description	❷ Initials	❸ Date
001			
002			
003			
004			
005			
006			
007			
008			
009			
010			
011			
012			
013			
014			
015			
016			
017			
018			
019			
020			
021			
022			
023			
024			
025			
026			
027			

Form E011 (Procedure EP-2-2)

Figure 15. Engineering Change Proposal Number Assignment Log Form

10.0 COST ANALYSIS REPORT NUMBER ASSIGNMENT LOG

10.1 Cost Analysis Report Number Assignment Log Form Preparation

The procedure for processing the Cost Analysis report Number Assignment Log form shall be followed by each individual responsible for entering information on the form. Each circled number below corresponds to the circled number on the Cost Analysis Report Number Assignment Log form E012. (See Figure 16. for an example of the form.)

10.1.1 Document Control

Obtain the following information from the Cost Analysis Report form E038. (Reference: EP-6-6, Cost Analysis Report.)

❶ Enter the part number and description.

❷ Enter your initials.

❸ Enter the date that the Cost Analysis Report number was assigned.

COST ANALYSIS REPORT NUMBER ASSIGNMENT LOG			
Cost Analysis Report No.	❶ Part Number and Description	❷ Initials	❸ Date
001			
002			
003			
004			
005			
006			
007			
008			
009			
010			
011			
012			
013			
014			
015			
016			
017			
018			
019			
020			
021			
022			
023			
024			
025			
026			
027			

Form E012 (Procedure EP-2-2)

Figure 16. Cost Analysis Report Number Assignment Log Form

11.0 PROTOTYPE REPORT NUMBER ASSIGNMENT LOG

11.1 Prototype Report Number Assignment Log Form Preparation

The procedure for processing the Prototype Report Number Assignment Log form shall be followed by each individual responsible for entering information on the form. Each circled number below corresponds to the circled number on the Prototype Report Number Assignment Log form E013. (See Figure 17. for an example of the form.)

11.1.1 Document Control

Obtain the following information from the Prototype Report form E039. (Reference: EP-6-7, Prototype Report.)

❶ Enter the part number and description.

❷ Enter your initials.

❸ Enter the date that the Prototype Report number was assigned.

PROTOTYPE REPORT NUMBER ASSIGNMENT LOG			
Prototype Report No.	❶ Part Number and Description	❷ Initials	❸ Date
001			
002			
003			
004			
005			
006			
007			
008			
009			
010			
011			
012			
013			
014			
015			
016			
017			
018			
019			
020			
021			
022			
023			
024			
025			
026			
027			

Form E013 (Procedure EP-2-2)

Figure 17. Prototype Report Number Assignment Log Form

12.0 BILL OF MATERIAL NUMBER ASSIGNMENT LOG

12.1 Bill of Material Number Assignment Log Form Preparation

The procedure for processing the Bill of Material Number Assignment Log form shall be followed by each individual responsible for entering information on the form. Each circled number below corresponds to the circled number on the Bill of Material Number Assignment Log form E014. (See Figure 18. for an example of the form.)

12.1.1 Document Control

Obtain the following information from the Bill of Material form E040. (Reference: EP-6-11, Bill of Material Change.)

❶ Enter the part number and description.

❷ Enter your initials.

❸ Enter the date that the Bill of Material number was assigned.

BILL OF MATERIAL NUMBER ASSIGNMENT LOG			
Bill of Matl. Report No.	❶ Part Number and Description	❷ Initials	❸ Date
001			
002			
003			
004			
005			
006			
007			
008			
009			
010			
011			
012			
013			
014			
015			
016			
017			
018			
019			
020			
021			
022			
023			
024			
025			
026			
027			

Form E014 (Procedure EP-2-2)

Figure 18. Bill of Material Number Assignment Log Form

13.0 CONFIGURATION BASELINE DOCUMENT NUMBER ASSIGNMENT LOG

13.1 Configuration Baseline Document Number Assignment Log Form Preparation

The procedure for processing the Configuration Baseline Document Number Assignment Log form shall be followed by each individual responsible for entering information on the form. Each circled number below corresponds to the circled number on the Configuration Baseline Document Number Assignment Log form E015. (See Figure 19. for an example of the form.)

13.1.1 Document Control

Obtain the following information from the Configuration Baseline Document form E002. (Reference: EP-1-8, Configuration Baseline Document.)

❶ Enter the description.

❷ Enter your initials.

❸ Enter the date that the Configuration Baseline Document number was assigned.

CONFIGURATION BASELINE DOCUMENT NUMBER ASSIGNMENT LOG			
CBD No.	❶ Description	❷ Initials	❸ Date
001			
002			
003			
004			
005			
006			
007			
008			
009			
010			
011			
012			
013			
014			
015			
016			
017			
018			
019			
020			
021			
022			
023			
024			
025			
026			
027			

Form E015 (Procedure EP-2-2)

Figure 19. Configuration Baseline Document Number Assignment Log Form

14.0 AS-BUILT CONFIGURATION RECORD NUMBER ASSIGNMENT LOG

14.1 Configuration Baseline Document Number Assignment Log Form Preparation

The procedure for processing the As-Built Configuration Record Number Assignment Log form shall be followed by each individual responsible for entering information on the form. Each circled number below corresponds to the circled number on the As-Built Configuration Record Number Assignment Log form E016. (See Figure 20. for an example of the form.)

14.1.1 Document Control

Obtain the following information from the As-Built Configuration Record form E003. (Reference: EP-1-8, As-Built Configuration Record.)

❶ Enter the description.

❷ Enter your initials.

❸ Enter the date that the As-Built Configuration Record number was assigned.

AS-BUILT CONFIGURATION RECORD NUMBER ASSIGNMENT LOG			
ABCR No.	❶ Description	❷ Initials	❸ Date
001			
002			
003			
004			
005			
006			
007			
008			
009			
010			
011			
012			
013			
014			
015			
016			
017			
018			
019			
020			
021			
022			
023			
024			
025			
026			
027			

Form E016 (Procedure EP-2-2)

Figure 20. As-Built Configuration Record Number Assignment Log Form

15.0 COMPONENT PART REQUEST NUMBER ASSIGNMENT LOG

15.1 Component Part Request Number Assignment Log Form Preparation

The procedure for processing the Component Part Request Number Assignment Log form shall be followed by each individual responsible for entering information on the form. Each circled number below corresponds to the circled number on the Component Part Request Number Assignment Log form E017. (See Figure 21. for an example of the form.)

15.1.1 Document Control

Obtain the following information from the Component Part Request form E028. (Reference: EP-3-9, Component Part Request.)

❶ Enter the part number and description.

❷ Enter your initials.

❸ Enter the date that the Component Part Request number was assigned.

COMPONENT PART REQUEST NUMBER ASSIGNMENT LOG			
CPR No.	❶ Part Number and Description	❷ Initials	❸ Date
001			
002			
003			
004			
005			
006			
007			
008			
009			
010			
011			
012			
013			
014			
015			
016			
017			
018			
019			
020			
021			
022			
023			
024			
025			
026			
027			

Form E017 (Procedure EP-2-2)

Figure 21. Component Part Request Number Assignment Log Form

16.0 REQUEST FOR ANALYSIS NUMBER ASSIGNMENT LOG

16.1 Request For Analysis Number Assignment Log Form Preparation

The procedure for processing the Request For Analysis Number Assignment Log form shall be followed by each individual responsible for entering information on the form. Each circled number below corresponds to the circled number on the Request For Analysis Number Assignment Log form E018. (See Figure 22. for an example of the form.)

16.1.1 Document Control

Obtain the following information from the Component Part Request form E036. (Reference: EP-6-3, Request For Analysis.)

❶ Enter the part number and description.

❷ Enter your initials.

❸ Enter the date that the Component Part Request number was assigned.

REQUEST FOR ANALYSIS NUMBER ASSIGNMENT LOG			
RFA No.	❶ Part Number and Description	❷ Initials	❸ Date
001			
002			
003			
004			
005			
006			
007			
008			
009			
010			
011			
012			
013			
014			
015			
016			
017			
018			
019			
020			
021			
022			
023			
024			
025			
026			
027			

Form E018 (Procedure EP-2-2)

Figure 22. Request For Analysis Number Assignment Log Form

17.0 ENGINEERING CHANGE NUMBER ASSIGNMENT LOG

17.1 Engineering Change Number Assignment Log Form Preparation

The procedure for processing the Engineering Change Number Assignment Log form shall be followed by each individual responsible for entering information on the form. Each circled number below corresponds to the circled number on the Engineering Change Number Assignment Log form E019. (See Figure 23. for an example of the form.)

17.1.1 Document Control

Obtain the following information from the Engineering Procedure form E037. (Reference: EP-6-4, Engineering Change Procedure.)

❶ Enter the part number and description.

❷ Enter your initials.

❸ Enter the date that the Component Part Request number was assigned.

ENGINEERING CHANGE NUMBER ASSIGNMENT LOG			
EC No.	❶ Part Number and Description	❷ Initials	❸ Date
001			
002			
003			
004			
005			
006			
007			
008			
009			
010			
011			
012			
013			
014			
015			
016			
017			
018			
019			
020			
021			
022			
023			
024			
025			
026			
027			

Form E019 (Procedure EP-2-2)

Figure 23. Engineering Change Number Assignment Log Form

18.0 DOCUMENT CHANGE NOTICE NUMBER ASSIGNMENT LOG

18.1 Document Change Notice Number Assignment Log Form Preparation

The procedure for processing the Document Change Notice Number Assignment Log form shall be followed by each individual responsible for entering information on the form. Each circled number below corresponds to the circled number on the Document Change Notice Number Assignment log form E020. (See Figure 24. for an example of the form.)

18.1.1 Document Control

Obtain the following information from the Document Change Notice form E041. (Reference: EP-6-12, Document Change Notice.)

❶ Enter the drawing number and description.

❷ Enter your initials.

❸ Enter the date that the Document Change Notice number was assigned.

DOCUMENT CHANGE NOTICE NUMBER ASSIGNMENT LOG			
DCN No.	❶ Drawing Number and Description	❷ Initials	❸ Date
001			
002			
003			
004			
005			
006			
007			
008			
009			
010			
011			
012			
013			
014			
015			
016			
017			
018			
019			
020			
021			
022			
023			
024			
025			
026			
027			

Form E020 (Procedure EP-2-2)

Figure 24. Document Change Notice Number Assignment Log Form

19.0 SPECIFICATION CHANGE NOTICE NUMBER ASSIGNMENT LOG

19.1 Specification Change Notice Number Assignment Log Form Preparation

The procedure for processing the Specification Change Notice Number Assignment Log form shall be followed by each individual responsible for entering information on the form. Each circled number below corresponds to the circled number on the Specification Change Notice Number Assignment Log form E021. (See Figure 25. for an example of the form.)

19.1.1 Document Control

Obtain the following information from the Specification Change Notice form E042. (Reference: EP-6-13, Specification Change Notice.)

❶ Enter the specification number and description.

❷ Enter your initials.

❸ Enter the date that the Specification Change Notice number was assigned.

SPECIFICATION CHANGE NOTICE NUMBER ASSIGNMENT LOG			
SCN No.	❶ Specification Number and Description	❷ Initials	❸ Date
001			
002			
003			
004			
005			
006			
007			
008			
009			
010			
011			
012			
013			
014			
015			
016			
017			
018			
019			
020			
021			
022			
023			
024			
025			
026			
027			

Form E021 (Procedure EP-2-2)

Figure 25. Specification Change Notice Number Assignment Log Form

20.0 EQUIVALENT ITEM AUTHORIZATION NUMBER ASSIGNMENT LOG

20.1 Equivalent Item Authorization Number Assignment Log Form Preparation

The procedure for processing the Equivalent Item Authorization Number Assignment Log form shall be followed by each individual responsible for entering information on the form. Each circled number below corresponds to the circled number on the Equivalent Item Authorization Number Assignment Log form E022. (See Figure 26. for an example of the form.)

20.1.1 Document Control

Obtain the following information from the Equivalent Item Authorization form E043. (Reference: EP-6-15, Equivalent Item Authorization.)

❶ Enter the part number and description.

❷ Enter your initials.

❸ Enter the date the Equivalent Item Authorization number was assigned.

EQUIVALENT ITEM AUTHORIZATION NUMBER ASSIGNMENT LOG			
EIA No.	❶ Part Number and Description	❷ Initials	❸ Date
001			
002			
003			
004			
005			
006			
007			
008			
009			
010			
011			
012			
013			
014			
015			
016			
017			
018			
019			
020			
021			
022			
023			
024			
025			
026			
027			

Form E022 (Procedure EP-2-2)

Figure 26. Equivalent Item Authorization Number Assignment Log Form

Title: **BLOCK NUMBER ASSIGNMENT**		Number: EP-2-3 Revision: A
Prepared by:	Approved by:	

1.0 PURPOSE

The purpose of this procedure is to define the method for assigning blocks of identification numbers.

2.0 APPLICABLE DOCUMENTS

Engineering Procedures:

EP-2-1, Document Numbering System

EP-2-2, Document Number Assignment Logbook

3.0 BLOCK NUMBER ASSIGNMENT PROCEDURE

3.1 Upon request from Engineering documentation groups, Document Control shall assign blocks of document numbers to specific product groups. Verification of need shall be established prior to block number assignment. If a request is made and verification of need cannot be proven, no document numbers will be assigned.

3.1.1 Verification of need consists of checking the Parts Database and confirming that previously assigned document numbers have all been utilized.

3.2 Assignment shall be made from the Document Number Assignment Logbook coinciding with the requestor's specific product group. Assignment of Document Numbers shall not exceed more than 2,000.

3.3 Assignment shall be made in such a manner so that any existing gaps in numeric order will be used-up. At such time as numeric gaps no longer exist, assignment of document number blocks shall proceed in sequential order.

3.4 Assignee shall record document number assignments in the proper document number assignment logbook.

3.5 Document Control will assign part number blocks to indicate general categories of parts and to conserve available document numbers.

3.6 Proposals for block number assignments will be initiated by the Engineering Change Analyst and approved by the Engineering Change Analyst. Block numbers will be issued by Document Control from a logbook.

Examples:

Document Type	**Number Range**
Product Specifications	PD001 thru PD999
Drawings	00001-00 thru 99999-99
Bill of Material	BM00001-00 thru BM99999-99
Specifications	SP00001 thru SP99999
Engineering Lab Notebooks	EL001 thru EL999
Reports	RP00001 thru RP99999

Title: **MODEL NUMBERING SYSTEM**		Number: EP-2-4 Revision: A
Prepared by:	Approved by:	

1.0 PURPOSE

The purpose of this procedure is to define the model number assignment method that is used to identify end products.

2.0 APPLICABLE DOCUMENTS

None.

3.0 GENERAL REQUIREMENTS

Engineering and Marketing work together to establish a model number that will capture the attention of customers. Also, Engineering will establish a location on the end product where the model number will be displayed to enhance the end product. The model number will not change as assemblies, subassemblies and parts are changed.

Model numbers do not apply to the Development and Preproduction phases. They shall be created for each end product that is released to Production.

4.0 MODEL NUMBERING PROCEDURE

An example of a Model number as it appears in documents is shown below:

Model Number
5000-XX

Base No. = 5000

Dash No. = XX

4.1 Base Number

The base number is always four digits long and begins with 5. The next three digits coincide with the first three digits of the Feature Code Number. If the Feature Code Number is 012 then the Model number would be 5012. (See Paragraph 4.3 for the Feature Code Number.)

4.2 Dash Numbers

As with Make-or-Buy parts, the Model dash number consists of two digits. However, it represents different customer options rather than the non-interchangeablity of one item to another. Therefore, several Model numbers may be active at once for a given end product number, as different options are available to customers at any given time.

4.3 Feature Code Number

When assigned to the first customer option offered for a given model, it coincides with the last two digits of the model number, e.g., for model 5012, the Feature Code dash number for the first option offered is 12. As new options are added, the next sequential dash number is assigned (but those which have already been assigned remain active).

4.4 Finished Goods Assembly Numbers

Finished goods assembly numbers are used to indicate the time at which a particular Top Level Assembly was active and when it was obsolete. One Finished Goods number exists for each model. Each Finished Goods number has a Bill of Material which calls out all of the various Top Level Assemblies for that model with their effective and obsolete dates.

Following is an example of the Model Numbers' and its associated engineering document numbering systems:

Model Number
Feature Code Number
Finished Goods Number
Top Level Assembly Number
Bill of Material

Title: **PART NUMBERING SYSTEM**		Number: EP-2-5 Revision: A
Prepared by:	Approved by:	

1.0 PURPOSE

The purpose of this procedure is to provide instructions, and to assign responsibilities for assigning new part numbers to items that are inventoried (parts, subassemblies or assemblies). It will also provide instructions for preparing and submitting the Part Number Request form.

2.0 APPLICABLE DOCUMENTS

Engineering Procedure:

EP-2-2 Document Number Assignment Logbook

Engineering Form:

E023, Part Number Request

3.0 DEFINITIONS

3.1 Buy Part/Assembly

A part or assembly which is made uniquely for the company by an outside vendor to the company's unique design specifications, that are supplied to the vendor in the form of engineering drawing(s) and/or engineering specification(s) accompanied by a bill of material (if an assembly). Printed circuit boards require additional documentation such as reproducible artwork that is supplied to the vendor that will be manufacturing the part.

3.2 Standard Part

An off-the-shelf item which is available commercially and is not made to any specifications other than those of the vendor.

3.3 Part Number

A number which uniquely identifies a piece part, subassembly or an assembly. For make-or-buy parts, the part number consists of the base number plus the dash number.

Example:

Base Number	**Dash Number**
00675	-02
Actual part number: 00675-02.	

3.4 Base Number

A five-digit number included in the part number of a make or buy part which identifies it as a unique design part. This number is assigned by the company to a new part, its engineering drawing and bill of material. It includes any zeros added to the left side of the number which are needed to make it a five-digit number.

3.5 Dash Number

The last two digits of a make or buy part number. The dash number changes whenever a non-interchangeable change is made to the part. The dash in a dash number shall appear in all documentation. When a new part number is assigned , the dash number shall always begin with -00.

3.6 Revision Level

3.6.1 The revision level is a one or two alpha character which represents changes to an engineering drawing. The revision level is not part of the part number. It changes each time the engineering drawing changes (whether the actual part changes or not). The main purpose of the revision level is to account for changes to the engineering drawing.

Example:

Base Number	Dash Number	Revision Level
00675	-02	D
Actual part number and document revision: 00675-02 D.		

3.6.2 Another purpose of the revision level, particularly at the assembly level, is to account for non-interchangeable changes to a part within a bill of material. Whenever a non-interchangeable change is made to a piece part, its assembly part number must either change or receive a revision level change.

4.0 GENERAL REQUIREMENTS

4.1.1 All part numbers for company parts used in production contain five digits. For those production part numbers which contain less than five digits, zeros must be added to the left of the part number to make it a five-digit number. This helps align the numbers for easy sorting.

4.1.2 Because any zeros added to the beginning of a company part number becomes an integral part of that part number, they shall appear on all company documentation representing that part number.

Note: *This applies to parts and assemblies used in production only. Specifications, product support manuals, etc. that have numbers assigned to them, do not follow this rule unless they are input to the company accounting system.*

4.1.3 All engineering and production part numbers shall be obtained from Document Control.

5.0 PART NUMBER PROCEDURE

This procedure is divided into the following parts:

- Part Number Procedure
- Part Number Request Form Preparation

The following procedure describes who is responsible and what they are supposed to do for each processing step.

5.1 Engineer

<u>Steps</u>

1. Prepare the Part Number Request form E023. (See Figure 27. for an example of the form.) Generate a new part number using the following numbering system:

2. Forward the completed Part Number Request form to the Engineering Manager for review and approval.

5.2 **Part Numbering System for a Typical Make-or-Buy Part.** The following is an example of a typical company make-or-buy part number and revision level as it appears in the title block of an engineering drawing.

Example:

Base Number	Dash Number	Revision Level
00221	-06	B
Actual part number and document revision: 00221-06 B.		

5.2.1 **Base Numbers.** The base number for a typical make or buy part is always 5 digits long. A base number for a part will always remain the same, regardless of the type of change. However, if the dash number changes more than 99 times, a new base number must be assigned.

5.2.2 The base number should not be confused with the drawing number. The drawing number is the same thing as the part number (the base number plus the dash number), except on assembly drawings from the top level. The entire part number shall appear in the engineering drawing title block.

The base number system outlined above applies to all three product phases.

- Development
- Preproduction
- Production

5.2.3 **Dash Numbers.** The dash number always consists of two numeric digits. Upon initial assignment of a new part number, the dash number is always -00.

5.3 Whenever a non-interchangeable change is made to a part, the dash number changes by one digit. This can happen up to 99 times, as outlined in Paragraph 5.2.1.

Example: -01, -02, -03 ... -99.

5.3.1 When an interchangeable change is made to the part, the dash number remains unchanged. This means that there can be two different looking parts that can be stored in the same bin, because they are interchangeable.

5.3.2 The dash number should not be confused with the engineering drawing revision level. The dash number is an integral part of the part number, whereas the revision level is not. The dash number and revision level do not change simultaneously with one another as changes are made to a engineering drawing, because the dash number only changes per a non-interchangeable change and the revision level changes every time the engineering drawing changes, regardless of the type of change.

5.3.3 The dash number system outlined above applies to all three product phases.

▪ Development ▪ Preproduction ▪ Production

5.4 **Document Revision Levels.** The document revision level consists of a one or two digit alphabetic representation. The document revision level is the only thing that must change every time a change is made to an engineering drawing.

5.4.1 The engineering drawing revision level appears in the engineering drawing revision block, in the top right hand corner, and at the bottom of the drawing next to the part number. It also appears in the Document Master List database for each part number.

5.4.2 The engineering drawing revision level is not part of the part number; however, it must be included with the part number when ordering parts. When the part number is ordered, the vendor should have the most current revision level of the engineering drawing to build to.

5.4.3 When a non-interchangeable change causes a part number to change the new part number shall be assigned the next revision level. A non-interchangeable change to the part number 23500-00 revision A would become 23500-01 revision B.

5.4.4 Engineering drawing revision levels as they apply to the different product phases are detailed below.

- **Development Phase**

During the Development phase, document revision levels shall be designated P1 (for Development release), P2, P3, etc.

- **Preproduction Phase**

When the engineering drawing is released into the Preproduction phase, the revision level shall become A (for Preproduction release), and subsequent changes shall be designated B, C, etc. The Development revision level history shall be erased from the drawing when it is released to Preproduction.

- **Production Phase**

When the engineering drawing is released to the Production phase, the revision level shall continue on as D, E, etc. with the next change after release. (The revision level shall not change at the time of production release.) The Preproduction revision level history shall remain on the drawing when it is released to production.

Note: *When releasing a part from one phase to another, the part number shall remain unchanged. Only the engineering drawing revision level will change at Preproduction release, as outlined above.*

5.5 Engineering Manager

3. Verify that the Part Number Request form is complete and correct. If approved, sign and date, then return the form to the Engineer for further processing.

5.6 Engineer

4. Forward the approved Part Number Request form to Document Control for the assignment of part numbers.

5.7 Document Control

5. Assign the next available part number(s) from the Document Number Assignment Logbook EP-2-2, under Paragraph 5.0, Part Number Assignment Log. Then enter the number in the Part Number Request form E023.

6. Forward a copy of the Part Number Request form to the Engineer.

7. Run copies, stamp, distribute, then file the original in the New Part Number Request File by part number.

PART NUMBER REQUEST		
Engineer: ❶		Date:
Engineering Manager: ❷		Date:
❸ Part Number	❹ Description	❺ Manufacturer's Part Number, If Applicable.

Form E023 (Procedure EP-2-5)

Figure 27. Part Number Request Form

6.0 PART NUMBER REQUEST FORM PREPARATION

The procedure for assigning new part numbers shall be followed by each individual responsible for entering information on the Part Number Request form. Each circled number below corresponds to the circled number on the Part Number Request form E023. (See Figure 27. for an example of the form.)

6.1 Engineer

❶ Enter your name and the date when the form is complete and ready for processing.

6.2 Engineering Manager

❷ Enter your name and the date, if approved.

6.3 Document Control

❸ Assign the next available part number from the Document Assignment Logbook EP-2-2, under Paragraph 5.0, Part Number Request Number Assignment Log. Then enter the number in the Part Number Request form E023.

6.4 Engineer

❹ Enter the part description.

❺ Enter the Manufacturer's part number, if this is a purchased part.

Title: **PART MARKING SYSTEM**		Number: EP-2-6 Revision: A
Prepared by:	Approved by:	

1.0 PURPOSE

The purpose of this procedure is to provide instructions, and to assign responsibilities for specifying the marking of part identification, serialization and other information on end item.

2.0 APPLICABLE DOCUMENTS

None.

3.0 DEFINITIONS

3.1 End Item

A part, assembly, unit, set or system that is the finished product or the prime level of an assembly.

3.2 Manufacturer

An individual, firm, company, or corporation engaged in the fabrication of finished or semi-finished products and whose name, registered trademark, or code identification is specified.

3.3 Package

The smallest enclosure into which an item(s) is placed for protection during storage or shipment.

3.4 Part

One piece, or two or more pieces joined together which are not normally subject to disassembly without destruction of its designed use. (Examples: Weldment, screw, computer chip.)

3.5 Part or Identifying Number

A marking applied to an item or its package for the purpose of identifying the item as a unique part or assembly.

3.6 Permanent Marking

A method of identification which will remain legible during the normal service life of an item.

3.7 Serial Number

A numeric code assigned to an item to differentiate it from any other item of the same part or identification number.

3.8 Set

A unit or units and necessary assemblies and parts connected or associated together to perform an operational function. Also used to denote a collection of parts, such as a "Printed Circuit Board."

3.9 System

A combination of parts, assemblies and sets joined together to perform an operational function.

3.10 Temporary Marking

A method of in-process identification which can be removed without defacing or damaging an item.

3.11 Unit

An assembly or any combination of parts, subassemblies and assemblies mounted together that are normally capable of independent operation in a variety of situations.

4.0 PART MARKING PROCEDURE

4.1 General Requirements

4.1.1 Each drawing of a part, assembly, unit, set or other item of supply shall specify marking, using an acceptable method that is not detrimental to the hardware or will not adversely affect its life, utility or function.

4.1.2 Materials. e.g.. powders. liquids, , glue, etc. which by their physical nature cannot be marked, will be identified by marking the container.

4.1.3 Part identification for components of subassemblies and assemblies which are not subject to disassembly or repair will be specified by temporary markings.

4.1.4 Drawings of parts which do not have a suitable surface or sufficient space for marking will specify tagging or marking the container.

4.1.5 Government and industry standard parts or vendor items are not reidentified, except as specified in the requirements for altered or selected items.

4.1.6 Supplier items controlled by specification or source control drawings are identified as specified in the drawing type requirements. Marking requirements are not specified by company standards, but will be controlled by reference to a Government standard. e.g., MIL-STD-130 or Industry standard, e.g., AS478.

4.2 MARKINGS

4.2.1 The height of permanent marking characters is established by the applicable standard and is not specified on the drawing, except when necessary to satisfy design or contractual requirements.

4.2.2 Whenever practicable, marking of a parts or assemblies shall be located so it is visible after installation.

4.2.3 Item numbers, used for reference purposes on the drawing, shall not be used as part identification numbers.

4.3 DRAWING APPLICATION

4.3.1 When design conditions permit, one or more optional methods of marking should be specified.

4.3.2 The marking process name is not specified on the drawing (electroetch, etc.).

4.3.3 Hardware cannot be identified with subsequent, part numbers. That is, a detail part cannot be marked with the assembly part number.

4.3.4 Parts are marked with one part number only.

4.3.4.1 In-process identification for parts is not normally specified on the engineering drawing; when necessary, it is specified by a temporary marking process that will not affect the surface to which the marking is applied and can be readily removed.

4.3.4.2 Parts containing more than one part number will show the difference between the numbers in the title. For example, an assembly containing a machined casting could have three part numbers; a casting part number, a machined part number and an assembly part number. The casting part number would have the prefix CSTG and the assembly part number would have the prefix "ASSY" to distinguish between the numbers.

4.3.5 Permanent part identification markings are normally specified on the field of the drawing, but may also be specified in a general note with a flag note ▷ referencing the location. The same note is used for either the field of the drawing or in the general notes.

4.3.6 When hardware is required to be permanently marked with a serial number, the requirement is specified by the addition of "AND ASSIGNED SERIAL NUMBER" to the marking note.

MARK PER (applicable standard) WITH (drawing number) AND APPLICABLE DASH NUMBER AND ASSIGN SERIAL NUMBER.

4.3.6.1 When an inseparable assembly contains a critical or major component that requires carry over serialization, the notation "AND SERIAL NUMBER OF "(applicable part number)" is added to the marking note.

MARK PER (applicable standard) WITH (drawing number and applicable dash number) AND SERIAL NUMBER OF (applicable part number).

4.3.7 Marking for additional requirements, such as quality control results, foundry trademark, forge shop symbol.

4.3.8 Matched sets or parts require markings to identify each component separately and as a part of the matched set. The complete matched set is assigned an assembly number and the detail part identification is correlated to the assembly part identification and serial number. The matched set is identified on the next assembly by the matched set assembly number. The part identification for each of the components comprising the matched set is specified by the following note:

MARK PER (applicable standard) WITH (applicable part number) PART NUMBER OR ASSY (applicable assembly number) AND ASSIGNED SERIAL NUMBER.

4.3.9 Altered and/or selected parts are reidentified with a company part number and the original identifying number is obliterated or removed. The requirement for the obliteration or removal is specified in the marking note, but the specific method or process to be used is left the discretion of the fabrication group.

REMOVE ORIGINAL PART NUMBER WITHOUT DAMAGE TO HARDWARE. MARK PER (applicable standard) WITH (applicable drawing number and applicable dash number).

4.3.10 Temporary markings are normally specified in a general note, but may be specified on the field of the drawing when necessary for clarity.

TEMPORARY MARKING PER (applicable specification) WITH (applicable drawing number and applicable dash number).

4.4 IDENTIFICATION PLATES, DECALS, LABELS AND OTHER MARKING DEVICES

4.4.1 The use of identification or information plates, decals, labels and other marking devices on assemblies, units, sets, systems or other end items is established by company policy and/or contractual requirements. It is the designer's responsibility to determine and specify the requirements for each project.

4.4.2 The identification plate or other marking device used is specified in the bill of material and on the field of the drawing. Attaching hardware or other requirements are also specified when necessary.

4.4.3 When markings are to be applied to the marking device, a note or detail and note will specify the markings and the method used.

4.4.4 The marking method selected shall produce markings that are as permanent as the expectancy of the device to which they are affixed.

4.4.5 When conditions will not permit marking the device in place, the marking note will indicate that the markings are to be affixed prior to assembly.

4.4.6 The marking device is not reidentified when markings are affixed at assembly.

Title: **SERIAL NUMBERING SYSTEM**		Number: EP-2-7 Revision: A
Prepared by:	Approved by:	

1.0 PURPOSE

The purpose of this procedure is to describe the serialization method used for traceability of parts, subassemblies, and final assemblies.

2.0 APPLICABLE DOCUMENTS

None.

3.0 DEFINITIONS

3.1 I

I, is the exempt code symbol that means that an item is exempt from traceability prior to receipt of material or items by Receiving Inspection.

3.2 IT

IT, is the code symbol signifying that an item must be traceable back thru all its processes to its origin.

3.3 Configuration Control

Is the method used to control the items' correct configuration in relation to the applicable specification, drawings and customer orders.

3.4 Identification

Is the method by which an item and its record are permanently associated by means of a part, lot or serial number.

3.5 Serial Numbers

Are numbers assigned to an item at the time the item receives its control beginning. Control will begin when an item begins to receive its configuration, such as subassemblies.

3.6 Lot Serialization

Is one serial number used to control a group of parts when the parts are of the same material, same heat number, received on the same order and the lot is subjected to a given process as a group.

4.0 RECORDS

4.1 To insure that records are permanently associated with an I or IT item, each part or assembly procured, fabricated, processed, or assembled will be assigned a serial number from a Logbook. All associated records (Shop Orders, Rejections, Inspection Records, etc.) and the item will retain this serial number throughout fabrication, assembly and stock or until such time the item and its records are incorporated into the next or parent assembly.

4.2 When an item is procured, fabricated or assembled to a given configuration, a Tag will be filled out completely and attached to the item. This tag contains information that establishes the exact configuration of a given part and serial numbered item. This tag is used in addition to the physical application of identification to the item.

4.3 The Traceability and Configuration Baseline Document is used to record listing of items being incorporated into an assembly. This record lists part numbers, serial numbers, rejection numbers, numbers of the person installing an item, etc. This form is assigned a serial number which readily associates all items in an assembly with one serial number. (Reference: EP-1-8, Configuration Baseline Document.)

5.0 STOCK TRACEABILITY

5.1 The Purchase Order Number will be affixed to each sheet or piece of raw material, or each container of bulk material (i.e., screws, powder, welding rod, etc.).

5.2 Material identity to the Purchase Order will be retained at all times on raw stock, cut-off, containers, bins, etc.

6.0 TRACEABILITY SERIAL NUMBERS

6.1 Application

Example Serial Number:

10024 A 99 10466

10024 = The Code number assigned to a delivered item which identifies the hardware within the system. This number is used only on delivered items not on details that are going into the next assembly .

A = Indicates the month of year that the part or assembly was manufactured. 'A' being January, etc.

99 = Indicates the year that the part or assembly was manufactured.

10466 = The Next available number assigned numerically to the part and assemblies being serialized.

6.2 The letter and seven digit numbers will appear on all parts or assemblies requiring traceability. Serial numbers are to be placed directly below the part number. Rejection numbers are to be placed directly below the part number and serial number.

Title: **ITEM MASTER INPUT**		Number: EP-2-8 Revision: A
Prepared by:	Approved by:	

1.0 PURPOSE

The purpose of this procedure is to define the method used for inputting new or revised part numbers into the Item Master database.

2.0 APPLICABLE DOCUMENTS

Engineering Form:

E024, Item Master Input

3.0 OVERVIEW

The Item Master Input form is initiated by the Engineering Change Analyst when an approved Release Notice is received from Engineering. The Item Master Input form is used by Engineering, Inventory Control, Production Control, Purchasing, and Accounting to enter new part information into the system files for which each department is responsible. The Item master Input travels to each department and ultimately returns to Document Control for filing.

4.0 ITEM MASTER INPUT PROCEDURE

This procedure is divided into the following parts:

- Item Master Input Procedure
- How to Prepare the Item Master Input form

The following procedure describes who is responsible and what they are supposed to do for each processing step.

4.1 Engineering Change Analyst

Steps

1. Prepare the Item master Input form E024 upon receipt of an approved Document Release Notice from. (See Figure 28. for an example of the form.)

2. Forward the completed Item master Input form to Inventory Control for processing.

4.2 Inventory Control

3. Enter the requested information as defined on the Item Master Input form.

4. Forward the completed Item master Input form to Production Control for processing.

Title: **ITEM MASTER INPUT**	Number: EP-2-8

4.3 Production Control

5. Enter the requested information as defined on the Item Master Input form.

6. Forward the completed Item Master Input form to Purchasing for processing.

4.4 Purchasing

7. Enter the requested information as defined on the Item Master Input form.

8. Forward the completed Item Master Input form to Accounting for final processing.

4.5 Accounting

9. Enter the requested information as defined on the Item Master Input form.

10. Forward the completed Item Master Input form to Document Control for filing.

4.6 Document Control

11. Copy, distribute and file the completed Item Master Input form.

ITEM MASTER INPUT

Engineering: ❶

Part Number .... ________
Revision Code .... ________
Description .... ________
Unit of Measure .... ________
Product Code .... ________
Engineering Change Number & Date .... ________

Inventory Control: ❷

Source Code .... ________
Spare Code .... ________
Planning Code .... ________
Order Quanity .... ________
ABC Code .... ________

Production Control: ❸

Manufacturing Lead Time .... ________
Yield Percent .... ________
Shrinkage Factor .... ________
Forcasted Usage .... ________

Purchasing: ❹

Commodity Code .... ________
Purchasing Lead Time .... ________
Buyer Code .... ________

Accounting: ❺

Labor Hours .... ________
Material Cost .... ________
Vendor Cost .... ________

Engineering: ❻	Date:
Manufacturing:	Date:
Purchasing:	Date:
Quality:	Date:

Form E024 (Procedure EP-2-8)

Figure 28. Item Master Input Form

5.0 ITEM MASTER INPUT FORM PREPARATION

The procedure for processing the Item Master Input form shall be followed by each individual responsible for entering information on the form. Each circled number below corresponds to the circled number on the Item Master Input form E024. (See Figure 28. for an example of the form.)

5.1 Engineering Change Analyst

 Enter:

The number of the part for which the form is issued.

The appropriate revision code, if applicable.

The part's description (title).

The unit by which the part is measured (usually ea.).

The code of the product line for which the part is to be used.

The Engineering Change number and date of the change.

5.2 Inventory Control

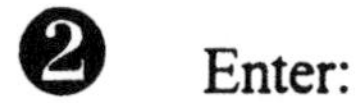 Enter:

The code that identifies the source of the part. A= Assembly, F= Fabricated, P= Purchased, N= Non-stocked Assembly, D= Document, R= Raw Material, X= Floor Stock, G= Finished goods.

The code that indicates whether this is a part to be used in maintenance or manufacture of a product. S= Spare Part, P= Product.

The code that indicates whether the part is planned or forecasted. U= Unplanned, P= Planned, F= Forecast/MRP, N= Forecast/No MRP

The estimated order size quantity.

The code for the part's Stockroom location.

5.3 Production Control

❸ Enter:

The number of working days from the time the kit is released from the Stockroom to the time the assembly should be completed.

The estimated yield percent from the manufacture of the part.

The estimated percent of the parts lost during manufacture.

The estimated forecasted usage.

5.4 Purchasing

❹ Enter:

The commodity code.

The number of working days between the time Purchasing receives a requisition for the part to the time that the goods are available for issue from the Stockroom.

The first and last initials of the buyer responsible for procuring the part.

5.5 Accounting

❺ Enter:

The standard labor hours required to produce this part.

The standard material cost of the part.

The outside labor is to be performed on the part, enter the standard cost; otherwise, leave blank.

5.6 All departments

❻ Enter:

Sign your name and date upon completion of the Item Master Input form.

Title: **BILL OF MATERIAL**		Number: EP-2-9 Revision: A
Prepared by:	Approved by:	

1.0 PURPOSE

The purpose of this procedure is to define the method used for preparing Bills of Materials. The Bill of Material will list the assemblies and components that are used to fabricate products.

2.0 APPLICABLE DOCUMENTS

Engineering Form:

E025, Bill of Material Input

3.0 OVERVIEW

A Bill of Material lists all of the components that will go into an assembled product and it will show each quantity required to make one assembly. These components can be subassemblies, purchased parts, manufactured parts, or raw materials. The Bill of Material Input form is initiated by the Engineering when the Bill of Material Product Structure file is created, and when changes to the current Bill of Material files are required.

4.0 BILL OF MATERIAL INPUT PROCEDURE

This procedure is divided into the following parts:

- Item Bill of Material Input Procedure
- How to Prepare the Bill of Material Input form

The following procedure describes who is responsible and what they are supposed to do for each processing step.

4.1 Engineering

Steps

1. Prepare the Bill of Material Input form E025 when the Bill of Material Product Structure file is created. (See Figure 29. for an example of the form.)

2. Forward the Bill of Material Input form to Manufacturing for processing.

4.2 Manufacturing

3. Review the Bill of Material Input form for completeness and accuracy.

4. If the Bill of Material Input form is not complete or is inaccurate, meet with Engineering to correct the form and approve the information recorded.

5. Forward the Bill of Material Input form to Engineering for processing.

4.3 Engineering

6. Forward the completed Bill of Material Input form to Document Control for filing.

4.4 Document Control

7. Copy, distribute and file the completed Bill of Material Input form.

BILL OF MATERIAL INPUT

Originator: ❶ Date:

Parent Part Number: Parent (assembly) Name:

Data Entry Code: ❷

- ❑ A to add a single level bill of material
- ❑ X to delete a single level bill of material
- ❑ B to add a component
- ❑ C to change an existing component's data
- ❑ D to delete a component

Item #❸	Comp. Part #	Qty Per	Eff. Start Date	Eff. Start Serial#	Eff. Close Date	Eff. Close Serial#	Del. to Oper Seq. #	Kit Lead Time Offset	Issuing Stores Acc.

Next Higher Assembly: ❹ Part Number:

Authorizations:

Engineering: ❺ Date:

Manufacturing: Date:

Form E025 (Procedure EP-2-9)

Figure 29. Bill of Material Input Form

5.0 BILL OF MATERIAL INPUT FORM PREPARATION

The procedure for processing the Bill of Material Input form shall be followed by each individual responsible for entering information on the form. Each circled number below corresponds to the circled number on the Bill of Material Input form E025. (See Figure 29. for an example of the form.)

5.1 Engineering

❶ Enter:

Your name and date that the form was completed.

The Parent Part Number and Description.

❷ Check the appropriate box to indicate the data entry activity to be performed.

❸ The item number that is recorded on the engineering drawing.

The component part number that is recorded on the engineering drawing.

The quantity per assembly that is recorded on the Engineering Change form

The effective start date that is recorded on the Engineering Change form.

The effective start serial number that is recorded on the Engineering Change form.

The effective close date that is recorded on the Engineering Change form.

The effective close serial number that is recorded on the Engineering Change form.

The deliver to operation sequence number. This information is obtained from the Manufacturing/Engineering manager.

The kit lead time offset figure in this field. This information is obtained from Materials.

The issuing stores account number in this field. This information is obtained from the Materials.

❹ Enter the next higher assembly and part number.

5.2 Engineering Manager

❺ Sign your name and date upon approval of the information entered on the form.

5.3 Manufacturing Manager

❻ Sign your name and date upon approval of the information entered on the form.

Title: **WHERE-USED INPUT**		Number: EP-2-10 Revision: A
Prepared by:	Approved by:	

1.0 PURPOSE

The purpose of this procedure is to define the method for determining which parts and/or assemblies are affected by an Engineering Change. It will also, define the departments responsible for completing this function and the areas in the change process where the 'Where-Used' look-up is required.

2.0 APPLICABLE DOCUMENTS

Engineering Form:

E026, Where-Used Input

3.0 OVERVIEW

3.1 A Where-Used check will always be performed on all Class 1 or 2 changes, and the Class 3 changes which are noted as 'Must Trace'. Engineering will be responsible for the accuracy of the Where-Used research and will perform the Where-Used look-ups when:

Processing an Engineering Change that changes an old part number to inactive.

Processing Class 1 or 2 Engineering Changes.

All part numbers that are changes because a change must have a Where-Used performed.

3.2 The originator of the an Engineering Change is responsible for obtaining a Where-Used look-up prior to starting a Change Control Board meeting. The Where-Used must be present at the Change Control Board meeting.

3.3 The Engineering Change Cover Sheet will be verified to insure that the products listed on the cover sheet are consistent with the results of the Where-Used in determining the Top Level (Product Types) that are affected by the change.

3.4 All old part numbers noted on the Where-Used reflect the product, model, feature/option and customer name of the next assemblies that may be affected by the change.

3.5 If all applications of the old part numbers are to be affected by the change then the Engineering Change Cover Sheet should be marked 'Yes'.

3.6 If all applications of all the old part numbers that are to be affected by the change are not impacted then the Engineering Change Cover Sheet will be marked 'No'. Justification for excluding certain part numbers from the change must be noted in the Where-Used by the originator of the Engineering Change.

3.7 The Bill of Material is the report that will be the source of most information on the Where-Used. Customer names will be obtained from the Engineering Change.

4.0 WHERE USED REPORT INPUT PROCEDURE

This procedure is divided into the following parts:

- Where Used Input Procedure
- How to Prepare the Where Used Input form

The following procedure describes who is responsible and what they are supposed to do for each processing step.

4.1 Engineering

Steps

1. Prepare the Where Used Input form E026 when there is an Engineering Change that affects the location of parts or assemblies within the product structure. (See Figure 30. for an example of the form.)

2. Forward the completed Where Used Input form to Manufacturing for processing.

4.2 Manufacturing

3. Review the Where Used Input form for completeness and accuracy.

4. If the Where Used Input form is not complete or is inaccurate, meet with Engineering to correct the form and approve the information recorded.

5. Forward the completed Where Used Input form to Engineering for processing.

4.3 Engineering

6. Forward the completed Where Used Input form to Document Control for filing.

4.4 Document Control

7. Copy, distribute and file the completed Where Used Input form.

WHERE USED INPUT

Originator: ❶ Date:

Engineering Change Number:

Old Part	Description	Qty Per	Next Assy	Description	Top Level P/N	Feature/ Option	Model No.	Product	Customer

Authorizations:

Engineering: ❷ Date:

Manufacturing: ❸ Date:

Form E026 (Procedure EP-2-10)

Figure 30. Where Used Input Form

5.0 WHERE USED INPUT FORM PREPARATION

The procedure for processing the Where Used Input form shall be followed by each individual responsible for entering information on the form. Each circled number below corresponds to the circled number on the Where Used Input form E026. (See Figure 30. for an example of the form.)

5.1 Engineering

Enter:

Your name and date that the Where Used Input form was completed.

The associated Engineering Change number.

The part number of the material currently being used that will be replaced, reworked, or scrapped by the proposed change.

A brief description of the old part number.

The assembly number(s) which the old part number is built into. The first level assembly into which the old part goes.

A brief description of the next assembly part number.

The first top level that appears for the next assembly(s). The top level part number is noted in the product structure tree (drawing) or configuration control document, basic units, features and options.

A brief description of the top level part number that identifies the unique feature/option.

The model number within the product family that identifies the product types.

The product family number that defines from which family the model type is.

The customer name whose product is affected by the change.

5.2 Engineering Manager

❷ Sign your name and date upon approval of the information entered on the form.

5.3 Manufacturing Manager

❸ Sign your name and date upon approval of the information entered on the form.

SECTION 3

DOCUMENTATION REQUIREMENTS

Title: **ENGINEERING DOCUMENTATION TYPES**		Number: EP-3-1 Revision: A
Prepared by:	Approved by:	

1.0 PURPOSE

The purpose of this procedure is to identify and describe the various engineering documents that are prepared to produce a product.

2.0 APPLICABLE DOCUMENTS

None.

3.0 GENERAL REQUIREMENTS

The task of explaining, specifying, and delineating the features of design, procurement, production, test, operation, and maintenance of a product requires the availability of many types of documents. Simple devices use only a few types. Complex systems require several types. Each type has been developed to resolve a specific situation. Selecting the right type reduces confusion and duplication. The document types discussed in this section are listed by category.

4.0 REFERENCE DOCUMENTS

4.1 General

Reference documents present technical information which is predominantly in written form. They may or may not include figures, charts, and illustrations. Reference documents are the verbal complement of pictorial data presented on drawings. Reference documents are used to explain details and give instructions which cannot be adequately documented on drawings.

4.2 Manuals

A manual describes a system and the functions it performs, relating to how the system is applied, serviced, operated, and maintained. Examples are: Theory of Operations Manuals, Users Manuals, and Field Service Manuals.

4.3 Procedures

A procedure is an instruction on how to do something, a particular way of accomplishing a task. A procedure is a series of steps which are followed in a definite order. Examples are installation procedures, test procedures, and assembly procedures.

4.4 Reference Specifications

Reference specifications are documents which are more global in nature than are engineering specifications. A plan or proposal may be a reference specification. Reference specifications are not revised nor controlled by Engineering Changes; therefore, they are not used to support actual specific items of hardware or software. Revisions to reference specifications must be carefully considered for their potential impact on existing methods and processes.

A standard is a generally agreed upon and accepted authority for how something should be in its final form. Documentation requirements are detailed in standards, as are many components whose use and design are more or less universal.

5.0 CONFIGURATION CONTROL DOCUMENTS

5.1 General

Configuration control and management documents define what is built, what is being built, and what will be built in the future. They also provide data for the location of specific systems to facilitate field retrofits and reworks.

5.2 Unit Product Code Index

The unit product code index lists all identifiers (product codes) which have been assigned to a particular model or system. Unit product codes are assigned to the major building blocks of a product to facilitate tracing of changes, reworks, and introduction of new items.

5.3 Machine Build Level

The machine build level controls the configuration of unit product code items in production. When a change or group of changes are ready for introduction into manufacturing, the manufacturing Bill of Material is updated and the machine build level is advanced. Engineering documentation is frequently at a higher level than what is being manufactured due to continuing development, Engineering Changes, and the economics of introducing changes in groups whenever practical.

5.4 As-Built Documentation

As-built documentation are copies of drawings of product elements which have been marked to show differences between drawing requirements and actual equipment configuration. As-built documents must be signed and certified as representing actual configuration.

5.5 Serial Number and Site

Serial number and site data is essential for tracking specific configurations which are supported by design engineering and field service. This data is maintained in a logistics folder, along with pertinent equipment unique data such as "as-built" documentation.

6.0 CHANGE DOCUMENTS

6.1 General

Change documents are used to facilitate the processes for revising documentation and subsequently changing the equipment being produced. Change documents establish release and control levels, permit outside functions to request engineering assistance and action, authorize changes by specifying points of introduction and disposition of parts, provide instructions for rework and provide tracking of previous configurations.

6.2 Document Release Notice

The Document Release Notice is a transmittal document which transfers physical control and tracking responsibility of specific documents from the originating group to Document Control. Levels of control, distribution requirements, and access restrictions are specified on the Document Release Notice form.

6.3 Engineering Change Request

The Engineering Change Request is a means by which persons may seek assistance or clarification, may point out errors or discrepancies, or may suggest improvements in design and methodology. Engineering change requests may be completed against released documents and hardware.

6.4 Engineering Change

The Engineering Change is the highest level of documentation control. For production released documentation, the Engineering Change is the sole authority and complete instruction for affecting changes to documents and hardware.

6.5 Rework Instructions

Rework Instructions are documents which are prepared to support an Engineering Change, when the nature of the rework is too extensive to be included on the Engineering Change. Multiple methods of rework may result in preparing separate rework instructions.

6.6 History File

The document history file is a repository of all revisions of each document controlled by Document Control. The documents filed are reproducible; they are used to investigate problems in down-level equipment and to support production until the next machine build level advance is completed.

7.0 FIELD DOCUMENTATION

7.1 General

Field documentation is data which supports Field Engineering in maintaining and servicing equipment after it has been placed with the customer. Field documentation is usually prepared by Field Engineering, with technical input from design engineering functions.

7.2 Field Bulletin

Field bulletins are issued to disseminate information and advise field personnel of potential products, technical tips, and new procedures. Field bulletins do not authorize changes, retrofit, or rework.

7.3 Field Instructions

The Field Instructions are the authority to install a change in equipment in the field. Field change notices emanate from Engineering Changes and are not issued until the parent engineering change is released.

7.4 List of Spares

The list of spares specifies all items which field engineering carries as spare parts. Knowledge of the contents of this list is essential to proper procurement and for adequate analysis of proposed Engineering Changes.

7.5 Compatibility Books

The Compatibility Book indicates which revision levels of equipment may be used with other various levels of equipment, without degradation to system performance, operation, or utility. The compatibility chart is particularly useful in (but not restricted to) the area of printed circuit board assemblies.

8.0 MANUFACTURING DOCUMENTS

8.1 General

Manufacturing documents are records of the special tooling, processes, operations, and methods required to produce and deliver the product delineated on engineering drawings. Manufacturing documents are generally internal in scope and application and are normally not placed under engineering change control.

8.2 Tooling

Tooling drawings depict the requirements for a unique apparatus which must be constructed or fabricated in order to facilitate the production of parts and assemblies. Examples are: assembly fixtures, templates, molds, and dies.

8.3 Process Specifications

Process specifications give detailed step-by-step instructions for successful achievement of a specific feature. Examples include assembly and test procedures, component loading or installation procedures, and wire routing processes.

8.4 Operations Sheets

Operations sheets are routing schedules used to direct in-process parts and assemblies through the production facility, in order to ensure all requirements are met in a timely and economical manner.

8.5 Shipping and Packing Documents

Shipping and packing documents specify the requirements for containers, packing materials, and packing and storage instructions for items ready for delivery. Several methods of shipping and packing may be required for a single item, depending upon Customs, Interstate Commerce, and Carrier regulations. Occasionally documents for shipping containers are placed under engineering change control in order to monitor color, logo usage, and cosmetic features of the package.

9.0 ENGINEERING DRAWINGS

9.1 General

Engineering drawings constitute the largest category of engineering documentation. This is due to the role technical drawings have enjoyed as the "language of industry", and to the almost universal understanding of the data they communicate. Most requirements may be placed directly on the product drawing. Data is transferred to auxiliary documents to avoid repetition, and to segregate areas of expertise.

9.2 Software Documents

Software is documented in the same general manner as hardware. This practice permits application of a single-change control procedure, provides a written record of revision history, and allows approvals to be readily obtained and indicated. The basic premise of software documentation is that listings, printouts, diskettes, disks, tapes, etc., are parts. Each unique item is assigned a dash number. All assigned dash numbers are listed on a document, along with the title, description, application, or other necessary data to assist in identifying the software element. The structure of software documentation is very simple because what is documented is source, i.e., the beginning elements and the system, after all the sources are linked, compiled, assembled, and made into a useable entity.

9.3 Source Documentation

Source documentation consists of a "cover sheet" type document which lists all identifiable elements of a specific file, and assigns a dash number to each element. A listing of each element is made, and its identifying number attached. This listing becomes a master reference document and is archived in the software section of Document Control. Source documentation is normally not distributed externally.

9.4 System Documentation

System Documentation consists of a "cover sheet" type document which lists the various source elements used to create the system. The linking, compiling, and assembly instructions may be included on the system drawing, or on another document, or on a media menu which is referenced on the system drawing. The media (disk, tape, etc.) delivered to users of the system is assigned a dash number, and identified in the same manner as component parts and assemblies are identified.

9.5 Hardware Drawings

Hardware drawings delineate the individual components which must be procured or fabricated and assembled into the end item. Part number identifications are established by hardware drawings. Noninterchangeable changes anywhere within a product's documentation structure will result in changing hardware drawings, because these changes are controlled by identifying part numbers.

9.6 Detail Drawings

A detail drawing depicts all necessary information for fabrication of an item (or family of items), ready for application in its using assembly. Several items within a family of sizes, values, etc., may be tabulated on a single detail drawing. Each unique part is assigned a separate dash number.

9.7 Assembly Drawings

An assembly drawing depicts the spatial relationship of two or more items required to form a part number identity of a higher order. There are two types of assemblies.

9.7.1 Inseparable Assembly

Inseparable assemblies are groups of items joined in such a manner that they may not be separated without destroying the components. Assemblies joined by welding, riveting, and fusing are inseparable assemblies. Inseparable assemblies are not supported by Bills of Materials. Line item numbers are assigned to each unique piece.

9.7.2 Separable Assemblies

Separable assemblies are groups if items which may be separated and rejoined without destruction. Separable assemblies are supported by Bills of Materials. Every piece of a production released separable assembly is assigned an part number. Assemblies joined by screws, bolts, pins, and retaining rings are separable assemblies.

9.8 Bills of Materials

Bills of Materials are the controlling element of all separable assemblies. In reality, an assembly drawing is a delineation of the data contained in the Bill of Material. Bills of Materials are prepared separately from assembly drawings. Bills of Materials may exist with no supporting assembly drawings. Bills of Materials carry the same document number and revision as their supporting assembly drawing. Every item required to complete the assembly is specified in the Bills of Materials.

9.9 Procurement Drawings

Procurement drawings are pre-approved for items provided by vendors and used in equipment that is designed and produced. Procurement drawings are prepared in order to:

1) Provide a part number for vendor items;
2) Specify which characteristics are essential for the item's fit and/or function in the equipment;
3) Establish approved sources for purchased items;
4) Record data for qualification of new sources;

5) Provide a formal interface between Engineering and the procurement process to more readily resolve problems of availability, lead time, cost, and vendor repeatability.

9.10 Approved Vendor List

An Approved Vendor List is prepared for each item depicted on commercial part and source control drawings. Approved Vendor Lists are prepared in order to:

1) Allow the addition or deletion of vendors without requiring an Engineering Change;

2) Serve as a source data input for updating the procurement database.

Approved Vendor Lists are not prepared for specification control or envelope drawings.

9.11 Commercial Part Drawings

Commercial part drawings are prepared whenever the source and the design characteristics of a vendor item must be controlled. Commercial part drawings depict known items of specified vendors. Multiple vendors are required for all commercial parts. Completed approved vendor lists are mandatory to support commercial part drawings and must accompany the initial release of the commercial part drawing.

9.12 Specification Control Drawings

Specification control drawings depict items whose design is established by existing specifications generated by industry, professional societies, or governmental agencies. Such items include screws, nuts, bolts, washers, resistors, capacitors, and light bulbs. Approved vendor lists are not required with specification control drawings. Items depicted on a specification control drawing are often produced by many vendors and are normally universally available. The actual source is not relevant to the suitability of the item.

9.13 Source Control Drawings

Source control drawings depict items whose design characteristics and requirements were developed (completely or in part) by Engineering. This type of drawing is used when Engineering has determined that only certain sources are qualified to produce the item because of the state-of-the-art tooling development, or predictability of critical requirements. Such items may include molded parts, crystals, power supplies, and cabinets whose internal structure is designed by the manufacturer.

9.14 Envelope Drawings

An envelope drawing depicts an item for which no vendor or commercially available part is known. This type of drawing sets forth certain design requirements to facilitate finding a commercially available part, or a vendor who will make and market such an item.

9.15 Support Drawings

Support drawings are prepared to complement hardware drawings. They often are not absolutely essential to the production of hardware, but serve as resource data for test, installation, start-up, maintenance, and definition of the hardware.

10.0 DETERMINING TYPES OF PROCUREMENT DRAWINGS

10.1 Film Masters

Film Masters are camera copies of original art work, such as printed circuitry, silkscreens, logos, or drawings. Film Masters are used to alleviate the problems of deterioration of most media used to prepare art work. Photoplots of printed circuits are documented as film masters.

10.2 Ship Groups

Ship groups are items "kitted" to accompany a product, option, or accessory. Ship groups contain materials and supplies which serve to assure the unit is functioning properly. Reference manuals, instructions, and log books are also included in ship groups.

10.3 Diagrams

Diagrams are graphic presentations using standard symbols, codes, and interconnecting lines to describe the function or interconnections of a system. Schematics, connection drawings, interconnect, and logics are diagrams.

10.4 Installation Drawings

Installation drawings depict how a system is arranged in the customer's facility. Data included consists of locations of power outlets, modifications required, and access space required.

11.0 ENGINEERING SPECIFICATIONS

Engineering Specifications define the requirements, operating environments, and/or the design characteristics of a material, part, assembly, sub-system, or system. Engineering specifications are assigned standard seven-digit document numbers, and may be released into any document release level. There are three basic types of engineering specifications. The intended application of each type is outline below.

11.1 Product Specification

A document which establishes the essential design features and operational characteristics of a part, assembly, or system. Product specifications are particularly useful for qualifying new sources for Original Equipment Manufacturer items or units procured commercially.

11.2 Material Specification

A document which defines the properties of a material to such extent as to preclude the selection and use of another material. Material specifications are used to eliminate ambiguities in the selection of such items as paint, printed circuit board materials, etc.

11.3 Process Specification

A document which gives instructions and inspection criteria for each step in a fabrication, assembly, or quality process. Process specifications are well-suited for acceptance test procedures, finish requirements, etc.

Title: **DOCUMENT FLOW CHART**		Number: EP-3-2 Revision: A
Prepared by:	Approved by:	

1.0 PURPOSE

The purpose of this procedure is to show the flow of engineering drawings and specifications.

2.0 APPLICABLE DOCUMENTS

None.

3.0 Engineering Drawings and Specifications

3.1 Engineering drawings constitute the largest category of engineering documentation. This is due to the role technical drawings have enjoyed as the "language of industry", and to the almost universal understanding of the data they communicate. Therefore, the drawing is shown as the principal document and how it flows, through the documentation system, for new drawing and revised drawings. (See Figure 31.)

3.2 The specification that is shown in the chart is actually the governing document with the drawing being a subordinate to the specification.

3.3 The Flow chart shows the associated documentation that is required to accomplish the documentation flow process.

Title: **DOCUMENT FLOW CHART**	Number: EP-3-2

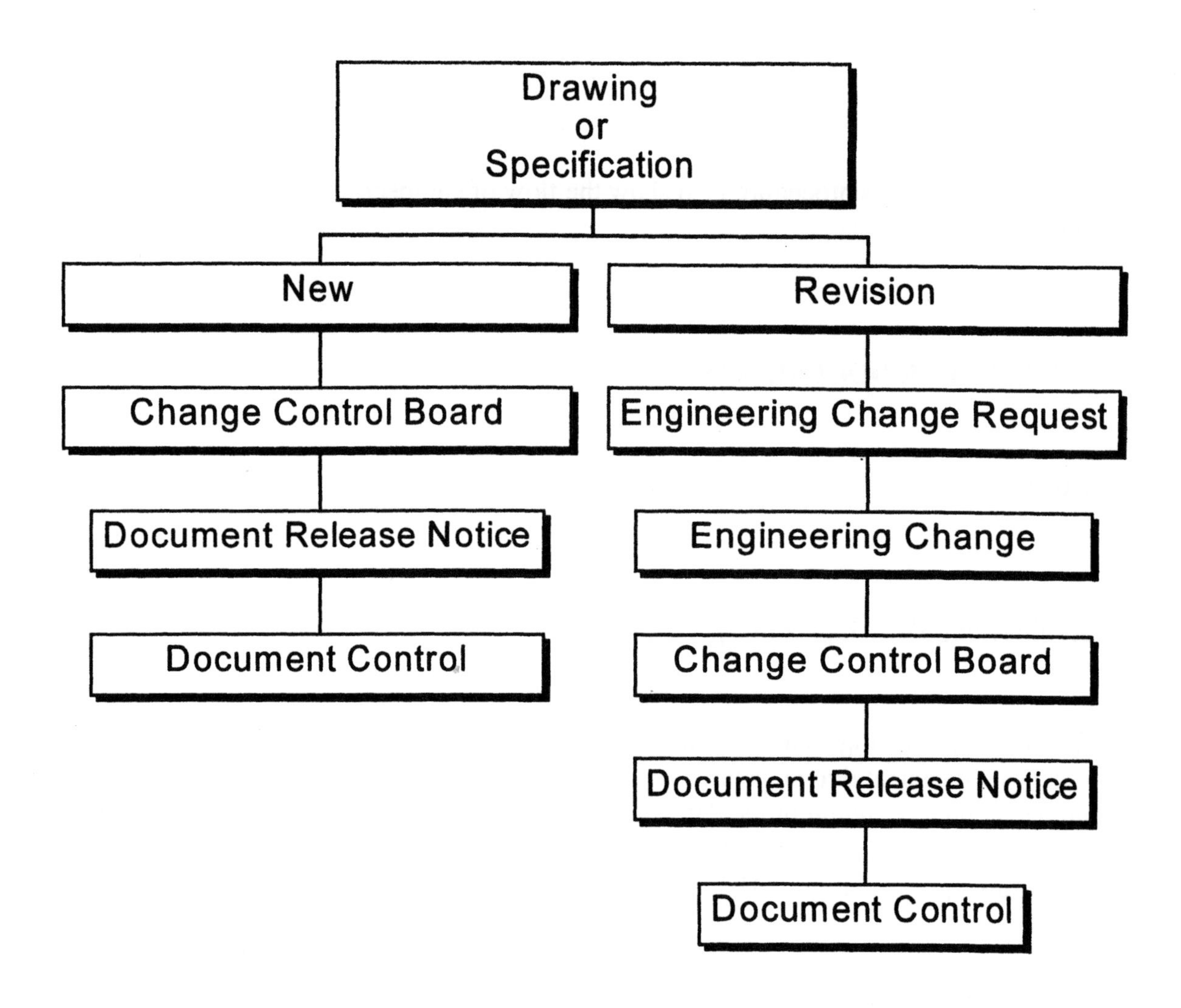

Figure 31. Document Flow Chart

Title: **DOCUMENTATION ORDER OF PRECEDENCE**		Number: EP-3-3 Revision: A
Prepared by:	Approved by:	

1.0 PURPOSE

The purpose of this procedure is to set forth the order of precedence to be followed for compliance with documents governing engineering designs and operations.

2.0 APPLICABLE DOCUMENTS

None.

3.0 OVERVIEW

Numerous documents exist which govern engineering design and the engineering business operation. This procedure establishes a hierarchy of priority for compliance with these documents. Documents governing design are addressed separately from those governing operations.

3.1 Order of precedence for engineering design:

1. Customer Contract
2. Engineering Design Standard and Drafting Standards
3. Functional Product Specification
4. Engineering Drawings and Specifications which define the product

5. Specifications and documents referenced within engineering drawings and specifications

3.2 Order of precedence for engineering operation:

1. Company Policies Manual
2. Company Standard Practices Manual
3. Company Engineering Procedures Manual
4. Company Divisional Standards and Practices Manuals
5. Company Department Instructions Manuals

Title: **ENGINEERING LAB NOTEBOOK**		Number: EP-3-4 Revision: A
Prepared by:	Approved by:	

1.0 PURPOSE

The purpose of this procedure is to provide instructions, and to assign responsibilities for the preparation of Engineering Lab Notebooks and their requirements. It will also provide instructions for preparing and submitting the Engineering Lab Notebook Request form.

2.0 APPLICABLE DOCUMENTS

Engineering Procedure:

EP-2-2, Document Number Assignment Logbook

Engineering Form:

E027, Engineering Lab Notebook Request

3.0 OVERVIEW

Engineering Lab Notebooks shall serve as the initial documentation of the conception of inventions, as well as a daily record of technical investigation, analysis, and findings. Data recorded in Engineering Lab Notebooks form the basis for product design and establishes a legal record of conception and witnessing of technical innovation. Specific requirements regarding Engineering Lab Notebooks are set forth in this document. Compliance is essential, as these notebooks often serve as evidence in legal action arising from patent disputes.

4.0 GENERAL REQUIREMENTS

All engineers involved in design and development activities must keep and maintain Engineering Lab Notebooks. Each engineer is responsible for the security of their notebooks. Engineering Lab Notebooks are restricted and confidential and must be treated as such.

While all engineers and technicians are required to comply with this standard as a condition of employment, Engineering Lab Notebooks are the property of the Company and, as such, must be surrendered to Document Control immediately upon termination.

Engineering Lab Notebooks are supplied, issued, numbered, and controlled by Document Control. The front cover of each notebook will have a non-removable label containing the following: department, name of responsible engineer or technician, company name and address, date allocated, and notebook control number.

5.0 ENGINEERING LAB NOTEBOOK CONTENTS

The first page shall be reserved for a table of contents. The top of each page shall be labeled briefly to aid in data organization and recall. The content shall include, but not be limited to:

- Design ideology
- Test methodology and results
- Engineering reasoning
- Engineering analysis (i.e. circuit, mechanical structure, detail part, system, etc.)
- Theoretical derivations behind engineering measurements

- Graphs and diagrams, as appropriate
- Competitor product data influential to product design
- Technical or reference documents which influence the design (including letters, vendor data, reports, and other supporting documents)
- Status of project, including records of presentations, design reviews, etc.

All information relating to potentially patentable ideas must be documented in an Engineering Lab Notebook and must be signed and dated by the inventor and a witness not involved in the subject of the entry.

Because it is often difficult to identify potentially patentable ideas in the initial stages, this process of signing and witnessing is recommended for all new entries on a regular basis.

Pages shall not be removed from Engineering Lab Notebooks.

Entries made in error shall not be erased or covered with correction fluid or tape. Errors shall be marked through with a single line.

When circumstances make it necessary, it is permissible to tape entries into the notebook from a separate sheet of paper.

Note: *Engineers should be aware that such entries do not provide as strong of a legal record. Entries which could be used for legal purposes shall be made directly into the notebook.*

All entries shall be made in ink, as markings by pencil are erasable.

Entries shall be made in chronological order as much as possible.

6.0 ENGINEERING LAB NOTEBOOK PROCEDURE

This procedure is divided into the following parts:

- Engineering Lab Notebook Procedure
- Engineering Lab Notebook Request Form Preparation

The following procedure describes who is responsible and what they are supposed to do for each processing step.

6.1 Originator

Steps

1. Prepare the Engineering Lab Notebook Request form E027 to request a new Engineering Lab Notebook. (See Figure 32. for an example of the form.)

2. Forward the completed Engineering Lab Notebook Request form to your manager for review and approval.

6.2 Manager

3. Verify that the Engineering Lab Notebook Request form is complete and correct. If approved, sign and date, then return the form to the Originator for further processing.

6.3 Originator

4. Forward the approved Engineering Lab Notebook Request form to Document Control for the assignment of an Engineering Lab Notebook number.

6.4 Document Control

5. Assign the next available number from the Document Number Assignment Logbook EP-2-2, under Paragraph 2.0 Engineering Lab Notebook Number Assignment Log. Then enter the number in the Engineering Lab Notebook Request form E027.

6. Forward a copy of the Engineering Lab Notebook Request form to the Originator along with a new Engineering Lab Notebook.

7. File the Original of the Engineering Lab Notebook Request form for future action when the Engineering Lab Notebook is returned.

6.5 Originator

8. Enter the following information on the Engineering lab Notebook label: department, name of responsible engineer or technician, company name and address, date allocated, and notebook control number.

9. Engineering Lab Notebooks shall remain locked up when not in use.

10. When the Engineering Lab Notebook is full, return it to Document Control for filing.

6.6 Document Control

11. Pull original Engineering Lab Notebook Request form and have the Originator enter their name and date under "Returned by". Then enter your name and date the form under "Received by".

ENGINEERING LAB NOTEBOOK REQUEST	
Requestor: Department No.: ❶ Date:	Notebook No.: ❷
Manager: Department No.: ❸ Date:	New ☐ Reassigned ☐ ❹
Notebook numbers currently assigned, if any: ❺	
Purpose of the notebook (program/description): ❻	
❼ I have read and understand Engineering Procedure EP-3-4 pertaining to the purpose for and use of Engineering Lab Notebooks. I agree to follow the instructions in the existing procedure, as well as any and all updates or changes to the existing procedure as they occur.	
Document Control Issued by: ❽ Date:	
Engineering Lab Notebook return: Returned by: ❾ Date:	
Document Control: Received by: ❿ Date:	

Form E027 (Procedure EP-3-4)

Figure 32. Engineering Lab Notebook Request Form

7.0 ENGINEERING LAB NOTEBOOK REQUEST FORM PREPARATION

The procedure for processing the Engineering Lab Notebook Request form shall be followed by each individual responsible for entering information on the form. Each circled number below corresponds to the circled number on the Engineering Lab Notebook Request form E027. (See Figure 32. for an example of the form.)

7.1 Originator

❶ Enter your name, date and Department Number .

7.2 Document Control

❷ Assign the next available number from the Document Number Assignment Logbook EP-2-2, under Paragraph 2.0 Engineering Lab Notebook Number Assignment Log. Then enter the number in the Engineering Lab Notebook Request form E027.

7.3 Manager

❸ Enter your name, date and Department Number.

7.4 Originator

❹ Check the appropriate box depending on whether this is a request for a new Engineering Lab Notebook or it is being reassigned to another engineer.

❺ Enter the number of each notebook that you have in your possession.

❻ Enter the purpose of the notebook.

❼ It is mandatory that the paragraph is read and understood.

7.5 Document Control

❽ Enter your name and the date that the Engineering Lab Notebook was issued.

7.6 Originator

❾ Enter your name and the date that the Engineering Lab Notebook was return.

7.7 Document Control

❿ Enter your name and the date that you received the Engineering Lab Notebook.

Title: **PRODUCT SPECIFICATION**		Number: EP-3-5 Revision: A
Prepared by:	Approved by:	

1.0 PURPOSE

The purpose of this procedure is to define the company standard for generating and formatting the product specification.

2.0 APPLICABLE DOCUMENTS

None.

3.0 DEFINITION

3.1 Product Specification

A descriptive document that depicts the physical appearance and functional characteristics of a product. Because this document contains the information required by a prospective customer to make a purchase decision, it serves as a work statement for Engineering, and becomes the bases for the formulation of marketing brochures and literature. It is also used by Engineering to formulate a complete product design.

4.0 PRODUCT SPECIFICATION PREPARATION

The product specification shall be the first document generated by Engineering to support a new product design. The product specification shell be coordinated with Marketing in order to provide correlation with the needs of the market place as determined by Marketing.

4.1 Titling of Specifications

The basic name of the product covered by the specification shall be the first part of the title. The title shall include, where appropriate, the minimum number of modifiers as are necessary for distinction and ready identification of the coverage of the specification. The type of specification "Product" shall be included above the specification title on the Cover Sheet only.

4.2 Contents

The content of the product specification shall include, but is not be limited to the following:

1.1 Introduction
1.1.1 Brief, introductory description of the product
1.2 Physical appearance
1.2.1 Size, dimensional aspects
1.3 Functional aspects
1.3.1 Capabilities
1.3.2 Limitations
1.3.3 Capacity
1.4 Operational aspects
1.4.1 Controls
1.4.2 Indicators
1.5 Features
1.5.1 Standards
1.5.2 Options

1.6 Environmental requirements and limitations
1.6.1 Temperature
1.6.2 Humidity
1.6.3 Altitude
1.7 Product safety
1.7.1 International standards with which product is to comply
1.8 Interface equipment
1.8.1 Requirements and limitations
1.9 Accessories
1.9.1 Special tools
1.9.2 Manuals
1.9.3 Diagnostic packages
1.10 Conversion kits
1.10.1 Speed conversions
1.10.2 Power conversions
1.10.3 Environmental conversions
1.11 Special customer requests
1.12 Compatibility with competition
1.12.1 Compliance with industry standards
1.12.2 Comparison to like products in the marketplace
1.13 Performance reliability
1.13.1 Quality standards which the product is to meet
1.13.2 Mean time between failures
1.14 Maintainability
1.14.1 Mean time to repair
1.14.2 Mean time between maintenance

4.3 Subject matter shall be kept within the scope of the above sections so that the same kind of information will always appear in the same section of every specification. If there is no information pertinent to a section, the following shall appear below the section heading:

"This section is not applicable to this specification."

4.4 Figures

A figure is any illustration, picture, schematic, or graph, and constitutes an integral part of the specification. It shall be clearly related to its associated paragraph. Figures shall be placed in numerical sequence at the end of the specification before any appendices. All figures shall be numbered consecutively with Arabic numbers in the order in which they are referenced in the specifications. All figures shall be titled, with the number and title appearing below the figure.

4.5 Tables

A table is used when data can be presented more clearly in lines and columns than in text. The placement, numbering, and titling of tables shall follow the same rules as for a figure except that the number and title of a table shall appear above the table.

4.6 Language Style

The most important consideration in a specification is its technical essence. This should be presented in language free of vague and ambiguous terms and using the simplest words and phrases that will convey the intended meaning. Sentences should be as short and concise as possible.

4.7 Abbreviations

Standard abbreviations may be used if they are repeated several times within the same specification. The first time an abbreviation is used in text, it shall be placed in parentheses following the word or term spelled out in full; i.e., pounds per square inch (psi). Abbreviations used in tables, figures, and equations shall be explained in the text or footnotes.

4.8 Commonly Used Words and Phrases

Certain words and phrases are frequently used in specifications. The following guidelines shall be used for such phrases.

4.8.1 Referenced documents shall be cited with the phrase "conforming to ...", "as specified in ...", or "in accordance with ...".

4.8.2 When making reference to a requirement in a specification, "as specified herein" shall be used if the requirement is not difficult to locate. Otherwise, the requirement shall be referenced by its section number.

4.8.3 "Shall" is used whenever a specification expresses a provision that is binding. "Should" and "may" are used wherever it is necessary to express non-mandatory provisions. "Will" may be used in cases where the simple future case is required, such as when expressing test results.

5.0 SECTIONAL ARRANGEMENT OF SPECIFICATIONS

5.1 Cover Sheet

The first page of a specification is the Cover Sheet. Centered in the main part of the form are the title of the specification and the current revision date. The footer, which appears on every page, contains the specification number, current revision level, title, and page _____ of _____.

5.2 Revision Sheet

The second page of a specification is the revision sheet. Required columns are: Revision, Engineering Change No., Section, Description, and Approval/Date.

5.2.1 Revision levels shall follow the same system as engineering drawings. During the Development phase, revision levels shall be designated P1 (for Prototype Release), P2, P3, etc. In the Preproduction phase, revision levels shall be designated "A" (for Preproduction Release), "B", "C", etc. Finally, revision levels in the Production phase shall take the next letter not used in the Preproduction phase and continue for as many letters as are needed.

5.3 The Engineering Change number is not required until the specification is released and until such time as its column may be filled in with "N/A". After the release of a specification, the applicable Engineering Change number must be filled in for any changes made to the specifications.

5.4 The section number(s) which contain changes per the current revision shall be listed separately on the revision sheet next to the description of the change. When the specification is being released during any phase, the section column should be filled in with "All".

5.5 Descriptions of changes shall follow the same format as engineering drawings.

5.6 The appropriate engineer shall initial and date the revision sheet in the approval/date column for the current revision only. For past revisions, the date is typed in on the word processor, and no approval is required.

5.7 When the specification is released, all Development revision history shall be removed from the revision sheet.

5.8 Table of Contents

If a specification contains seven (7) or more pages, a Table of Contents shall be included on a separate page immediately following the revision sheet. The Table of Contents shall list all major sections and subsections of the specification and the pages on which they are found.

5.8.1 Section 1 - Scope

A statement of the scope shall consist of a clear, concise abstract of the coverage of the specification. This brief statement shall be sufficiently complete and comprehensive to describe in general terms the item, material, or process covered by the specification in terms that may be easily interpreted by manufacturers, contractors, suppliers, etc.

5.8.2 Section 2 - Applicable Documents

All documents referenced elsewhere in the specification shall be listed in Section 2. Examples of such documents are Military Specifications and Standards, ANSI Standards, Engineering drawings, and other specifications. When listing documents in Section 2, they shall be grouped together by document type.

5.8.3 Section 3 - Requirements

The essential requirements and descriptions that apply to performance, design, reliability, etc. of the item covered by the specification shall be stated in this section. Examples of the type of information that should appear in this section are performance characteristics, physical characteristics, reliability, maintainability, environmental conditions, and materials. This section is intended to indicate the minimum requirements that an item must meet to be acceptable. The Requirements section shall be written so that compliance with all requirements will assure the suitability of the item for its intended purpose, and non-compliance with any requirement will indicate unsuitability for the intended purpose.

5.8.4 Section 4 - Quality Provisions

This section shall include all of the examinations and tests to be performed in order to assure that the product or item conforms to the requirements in Sections 3 and 5 of the specification. The test and inspection philosophy shall be described with a statement of responsibility for inspection, classification of examinations and tests, sampling, and other information pertinent to the quality provisions. All tests and inspections required shall be described in detail and shall state the minimum end results that the item must achieve to be acceptable.

5.8.5 Section 5 - Preparation for Delivery

This section shall include applicable requirements for preservation, packaging, and packing the item, and for marking of packages and containers.

5.8.6 Appendix

An appendix may be used to append large data tables, plans pertinent to the submittal of the item, management plans pertinent to the subject of the specification, etc. An appendix is identified by the heading "APPENDIX" and shall be referenced in the body of the specification. Use of an appendix is optional. If more than one appendix is used, they shall be numbered in the same manner as figures and tables.

Title: **CONFIGURATION CONTROL DOCUMENT**		Number: EP-3-6 Revision: A
Prepared by:	Approved by:	

1.0 PURPOSE

The purpose of this procedure is to define each configuration of an end product which has been placed on the Master Schedule. The Configuration Control Document is used to list each feature or option that is available for each product being produced. It can be used by Marketing to determine which features/options are available to customers at any given time, and what the current revision level is for each model.

2.0 APPLICABLE DOCUMENTS

None.

3.0 OVERVIEW

The Configuration Control Document is a document which lists each top level assembly part number (and other pertinent information) available for each model being produced. Its purpose is to indicate which Top Level Assemblies exist for each model, and the revision level history for each Top Level assembly. It can also be used by Reconditioning to identify the correct level of parts in a given product needing rework.

The Configuration Control Document plays a vital role in configuration control. Indication of the Top Level Assembly revision level on every product's part number label and use of the Configuration Control Document as a guide to the parts affected by each Top Level Assembly revision level change together assist in the proper identification of parts within each End Item.

4.0 GENERAL REQUIREMENTS

The Configuration Control Document is generated and maintained by Engineering. It is under Engineering change control and may therefore only be changed via an Engineering Change, except for the update of the implementation date for each Top Level Assembly revision.

4.1 One Configuration Control Document exists for each series of models.

4.2 The Configuration Control Document for each series of models shall be released at the time of Preproduction release of the first model within that series. As new features and models within a series are released, the Configuration Control Document shall be revised accordingly.

4.3 The Configuration Control Document shall change every time a non-interchangeable change is made to a part within the product structure.

4.4 The Configuration Control Document is intended for internal use only. Under no circumstances should it be given to customers for the purpose of choosing product features.

5.0 CONFIGURATION CONTROL DOCUMENT PROCEDURE

5.1 The Configuration Control Document shall be considered an Engineering Specification. Its cover page and revision page shall be prepared in accordance with and follow the same format as Specifications. The exception to this shall be that the column on the revision page titled "Section" shall be eliminated, as the Configuration Control Document contains no section numbers.

5.2 Finished Goods Assembly Number

One finished goods number exists for each model. The finished goods number allows for the grouping together of different Top Level Assemblies for a model, and for obsolescence of Top Level Assemblies when necessary. Finished goods numbers are assigned by the engineering change analyst.

5.3 Option

An option is a configuration available to a customer at the Top Level Assembly level. Usually the first option for a model is a product with a standard part. When customers request special parts in their product such as unique part or Printed Circuit Boards, new options are created. Creation and obsolescence of options is controlled by Marketing and Engineering.

5.4 Top Level Assembly Number

One Top Level Assembly number exists for each option available for a given model. Top Level Assembly variations represent the non-interchangeablity between parts at the Top Level Assembly level only.

5.5 Top Level Assembly Revision Level

This section lists the revision level changes which have occurred on each Top Level Assembly number. When a non-interchangeable change is made anywhere within the product structure of any Top Level Assembly, the revision level of the Top Level Assembly shall change.

5.6 Engineering Change Number

This section indicates the Engineering Change number which resulted in the Top Level Assembly revision level.

5.7 Part Affected

This section lists the part or parts which were affected by the non-interchangeable change for a revision, giving a quick reference for areas such as Reconditioning.

5.8 Engineering Change Release Date

This section indicates the date on which an Engineering Change which caused a revision to a Top Level Assembly was released. This date is not necessarily the same as the implementation date described in Paragraph 5.9.

5.9 Implementation Date

This section indicates the date on which a Top Level Assembly revision level (and Engineering Change number) was implemented on the manufacturing line. Because this date cannot be precisely determined until the change is actually implemented, it is blank at the time of Engineering Change release.

The Configuration Control Document shall be updated upon Engineering Change release without the implementation date. When the implementation date has been determined, Production Control shall supply this information to Document Control, and the Configuration Control Document shall be updated and redistributed (without an Engineering Change) with the implementation date filled in.

Title: **DRAWING FORMAT REQUIREMENTS**		Number: EP-3-7 Revision: A
Prepared by:	Approved by:	

1.0 PURPOSE

The purpose of this procedure is to define the company standard for engineering drawing format sizes.

2.0 APPLICABLE DOCUMENTS

None.

3.0 OVERVIEW

3.1 Standardization of drawing sizes and the location of format features on drawing forms provides a definite advantages in readability, handling, filing, and reproduction. In using drawings made by other organizations an advantage is gained when like items of information are in the same location on all drawings and when uniformity of form and language is applied in marking information entries.

3.2 Revision information and dates are of particular importance to users of drawings and should be located and expressed uniformly on all engineering drawings.

4.0 DRAWING SIZE AND FORMAT

4.1 Engineering drawings shall be prepared on standard preprinted drawing forms that are listed in the following Table.

4.2 The smallest size consistent with the drawing content and potential revisions shall be used.

4.3 Only those drawing sizes listed in the following Table shall be used.

4.4 The size designation appears in the title block of all preprinted drawing forms. (Sizes A thru E.)

4.5 All sheets of multisheet drawings shall be the same size.

Size Designation	Dimensions
A	8 1/2" X 11" (Vertical)
A	11" X 8 1/2" (Horizontal)
B	11" X 17"
C	17" X 22"
D	22" X 34"
E	34" X 44"

Title: **PRINTED CIRCUIT BOARD DOCUMENTATION**		Number: EP-3-8 Revision: A
Prepared by:	Approved by:	

1.0 PURPOSE

The purpose of this procedure is to define the engineering documentation required for the production of printed circuit boards.

2.0 APPLICABLE DOCUMENTS

None.

3.0 REQUIRED DOCUMENTS FOR PRINTED CIRCUIT BOARDS

3.1 General Requirements for Master Artwork

1. The master artwork may be prepared as a 1:1 photo positive by CAD equipment or it may be hand taped at an enlarged scale.

2. The artwork base material shall be a dimensionally stable material with a minimum thickness of .005 inches.

3. The artwork shall have shear marks to indicate the major corners of the board. The board edges shall be along the inside edges of the marks.

4. The artwork shall have three registration targets on each layer of the artwork. Targets shall be located off the horizontal and vertical edges of the board on .100 inch grid increments and spaced an exact distance apart to visually indicate the extent that the artwork is to be reduced.

5. The artwork shall have locations for three tooling holes located as far apart as possible.

3.2 Pad Master Drawing Requirements

1. The Pad Master consists of all pads, the shear marks, and reduction labeling between two targets.

2. The pad pattern should be labeled accordingly.

3.3 Component Traces Pattern Requirements

1. The component traces pattern consists of traces, running in one major direction, and displays the company "logo" or name, and "Made in USA".

2. The component traces pattern is labeled "component side".

3.4 Noncomponent Traces Pattern Requirements

1. The noncomponent traces pattern consists of traces, running in the other major direction, and displays the printed circuit board number and revision (PCB XXXXX Rev. A), and a place for the printed circuit assembly number and revision (PCA______ Rev. ___). The symbol Δ is used to indicate where the manufacturer is to place the UL approved mark.

2. The noncomponent traces pattern is labeled "circuit side".

3.5 Silk Screen Drawing Requirements

1. The silkscreen pattern consists of component outlines and reference designators which are placed outside of the component outline so as to be visible after the component is installed.

   Reference designators are assigned to the board from left to right and from top to bottom. When a series of boards are designed for one product, each board carries a unique series of reference designators. For example, the reference designators for one board would all start with a "1" - (1R1, 1C12 etc.), and the designators for the next board would all start with a "2" - (2R1, 2C12, etc.).

2. The first and last number of any connector and/or board edge contact is labeled on the silkscreen. The component polarity or orientation pin, tab, or mark on all components (when applicable) shall be indicated on the silkscreen.

3. The minimum letter size to be used on the silkscreen is .070 (scale 1:1).

4. The silkscreen has reduction labeling between two targets and is labeled "silkscreen".

3.6 Fabrication Drawing Requirements

1. The fabrication drawing of the printed circuit board shall depict the dimensional configuration of the board, size, and location of holes; and shell specify the material, process specifications, notes, and other information necessary for the fabrication of the printed circuit board. The fabrication drawing should show the circuit side of the board.

2. The three tooling holes, located as far apart as possible on the board, are the master reference points, with the exception of key interconnecting features, for all component hole locations and board dimensions. The tooling holes are located on grid centers and are .125 ±.002 inches in diameter unplated. The acceptable tolerance on tooling hole position is ±.003 inches.

3. The hole schedule chart, location in the lower right side of the drawing, contains the symbol for each hole, the diameter and tolerance of each hole, the quantity of each hole, and indicates whether or not each hole is placed.

4. Notes to appear on the fabrication drawing:

   Hole sizes are after plating.
   Holes located on _____ grid.
   Circuit side of board shown.
   Base laminate: FR4 -.062, .093, or .125 thick.
   Conductor: 1 oz copper minimum before plating-
   double sided or (single sided).
   Solder plating: 40% ±5% lead & 60% ±5% tin.
   Minimum plating thickness on trace surface and through holes: cooper - .001; tin-lead -.0003.

Minimum plating thickness for fingers: gold -.00005.
Solder mask material: must conform to IPC-SM-840A Class III glossy finish, using a .020 expanded pad master.
SMOBC/HASL (Solder mask over bare cooper, hot air solder level).
Silkscreen to be nonconductive yellow ink.
△ Add manufacturers emblem or name and UL#.

5. The fabrication drawing contains the engineering change history and is always sheet 1 of the printed circuit layout.

3.7 Solder Mask Drawing Requirements

1. The solder mask is created photographically from the past pattern such that the solder mask diameters exceed the pad diameters by .020 inches.

Note: *The pad pattern, the component traces pattern, the noncomponent traces pattern, the silkscreen pattern and the fabrication drawing are all considered to be one document.*

Title: **COMPONENT PART REQUEST**		Number: EP-3-9 Revision: A
Prepared by:	Approved by:	

1.0 PURPOSE

The purpose of this procedure is to explain how to introduce new or changed component parts to the existing Master File Database. It will also provide instructions for preparing and submitting the Component Part Request form.

2.0 APPLICABLE DOCUMENTS

Engineering Procedure:

EP-2-2, Document Number Assignment Logbook

Engineering Form:

E028, Component Part Request

3.0 GENERAL REQUIREMENTS

3.1 The Component Part Request form is initiated by any employee to introduce a new purchased part or to change manufacturer information for an existing purchased part. The Component Part Request, with complete manufacturer's information for the part, is submitted to the Engineering for review.

3.2 A Component Part Request with only one vendor source must be accompanied by an approved Single Source Authorization form before it will be reviewed by Engineering.

4.0 COMPONENT PART REQUEST PROCEDURE

This procedure is divided into the following parts:

- Component Part Request
- Component Part Request Form Preparation

The following procedure describes who is responsible and what they are supposed to do for each processing step.

4.1 Originator

Steps

1. Prepare the Component Part Request form E028 to introduce new or changed component parts to the existing Master File database. (See Figure 33. for an example of the form.)

2. Forward the completed Component Part Request form to your manager for review and approval.

4.2 Manager

3. Verify that the Component Part Request form is complete and correct. If approved, sign and date, then return the form to the Originator for further processing.

4.3 Originator

4. Forward the approved Component Part Request form to Document Control for the assignment of a Component Part Request number.

4.4 Document Control

5. Assign the next available number from the Document Number Assignment Logbook EP-2-2, under Paragraph 15.0, Component Part Request Number Assignment Log. Then enter the number in the Component Part Request form E028.

6. Forward a copy of the Component Part Request form to the Originator.

COMPONENT PART REQUEST	
Originator: Department No.: ❶ Date:	CPR No.: ❷
Manager: Department No.: ❸ Date:	Where Used: ❹
Reason for Request: ❺ ❑ New Component ❑ Alternate Source ❑ Investigation ❑ Other	
Description: ❻	
Manufacturer Name: Manufacturer Part Number: Manufacturer Name: Manufacturer Part Number: Note: New components require two sources.	
Cost (Estimate)	Project No.:
Action: ❑ Samples Requested Date: ❑ Qualification Test	
Remarks: ❼	
Authorization: Engineering: ❽ Date:	
Document Control: ❾ Date:	

Form E028 (Procedure EP-3-9)

Figure 33. Component Part Request Form

5.0 COMPONENT PART REQUEST FORM PREPARATION

The procedure for processing the Component Part Request form shall be followed by each individual responsible for entering information on the form. Each circled number below corresponds to the circled number on the Component Part Request form E028. (See Figure 33. for an example of the form.)

5.1 Originator

❶ Enter your Name, Department Number and the Date.

5.2 Document Control

❷ Assign the next available part number from the Document Number Assignment Logbook EP-2-2, under Paragraph 15.0, Component Part Request Number Assignment Log. Then enter the number in the Component Part Request form E028.

5.3 Manager

❸ Enter your Name, Department Number and Date, upon approval.

5.4 Originator

❹ Enter the numbers of all assemblies in which the part is used.

❺ New Component. Enter an X if requesting a new part.

Alternate Source. Enter an X if requesting that a new vendor be added to the Approved Vendor List.

Investigation. Enter an X if requesting that an investigation be performed on the component.

Other. Enter an X if none of the above options describes the reason for the request. Specify the reason in the space below.

❻ Enter the description of the part.

Enter the name of the manufacturer who makes the part.

Enter the manufacturer's part number.

Note: *For all new parts, two Manufacturers must be listed unless a Single Source Authorization is approved.*

Enter the name of the second manufacturer who makes the part.

Enter the second manufacturer's part number.

If requesting a new part, enter the estimated unit cost of the part. If requesting an alternate source, enter the estimated cost savings, if applicable.

Enter the number of the project or product in which the part is or will be used.

Enter an X after ordering samples of the part from the vendor.

Enter an X after qualification test if it has been performed. Enter the results of the qualification testing in the remarks section.

❼ Use this section to indicate the test results, the final disposition of the Component Part Request, and any other explanatory remarks.

❽ Engineering signs here to indicate final approval of the Component Part Request.

5.5 Document Control

❾ Enter your name and date when the Component Part Request is complete and ready for filing.

Title: **DELIVERABLE DOCUMENTATION REVIEW**		Number: EP-3-10 Revision: A
Prepared by:	Approved by:	

1.0 PURPOSE

The purpose of this procedure is to describe a system for verifying the completeness and correctness of deliverable documentation. It will also provide instructions for preparing and submitting the Deliverable Documentation Checklist form.

2.0 APPLICABLE DOCUMENTS

Engineering Form:

E029, Deliverable Documentation Checklist

3.0 GENERAL REQUIREMENTS

The responsibility of verifying the completeness and correctness of deliverable documentation is that of the Final Inspection department.

This work is to be done in conjunction with final acceptance processing.

Applicable contracts, purchase orders, and other agreements are to be examined to determine the documentation that is to accompany each deliverable item. This information is to be recorded on the Deliverable Documentation Checklist form.

4.0 DELIVERABLE DOCUMENTATION REVIEW PROCEDURE

This procedure is divided into the following parts:

- Deliverable Documentation Review
- Deliverable Documentation Checklist Form Preparation

The following procedure describes who is responsible and what they are supposed to do for each processing step.

4.1 Originator

Steps

1. Prepare the Deliverable Documentation Checklist form E029 to verify that all the deliverable documentation is in the package. (See Figure 34. for an example of the form.)

2. Forward the completed Deliverable Documentation Checklist form to your manager for review and approval.

4.2 Manager

3. Verify that the Deliverable Documentation Checklist form is complete and correct. If approved, sign and date, then return the form to the Originator for further processing.

4.3 Originator

4. Forward completed copy of the Deliverable Documentation Checklist form to the Final Inspection for follow-up.

4.4 Final Inspection

5. Individual items of documentation are to be examined to see if they are complete and, when applicable, apply to specifications for that type of documentation.

6. Forward the completed Deliverable Documentation Checklist form to the originator for filing.

DELIVERABLE DOCUMENTATION CHECKLIST			
Part Name: ❶		Part Number: ❷	Serial Number: ❸
Customer: ❹		PO or Contract: ❺	Date: ❶
Item Number	Documentation Item	Required ❻	Available ❼
1	Serialized Bill of Material		
2	As Built Drawings		
3	Spares List		
4	Operating Instructions		
5	Maintenance Instructions		
6	Final Inspection Report		
7	Unit Test Report		
8	System Test Report		
9	Hardness Tests		
10	Stress Test		
11	Weight Record		
12	Capacity Affidavit		
13	Certifications		
14	Nonconformance Reports		
15	Other (Specify)		
16	Other		
Remarks: ❽			
Prepared By: ❾		Date:	
Approved By: ❿		Date:	

Form E029 (Procedure EP-3-10)

Figure 34. Deliverable Documentation Checklist Form

5.0 DELIVERABLE DOCUMENTATION CHECKLIST FORM PREPARATION

The procedure for processing the Deliverable Documentation Checklist form shall be followed by each individual responsible for entering information on the form. Each circled number below corresponds to the circled number on the Deliverable Documentation Checklist form E029. (See Figure 34. for an example of the form.)

5.1 Originator

❶ Enter your Name and Date.

❷ Enter the Part number.

❸ Enter the Serial Number.

❹ Enter the customers name.

❺ Enter the Purchase Order number or the Contract Number.

5.2 Final Inspection

❻ Check the items that are required for the end item.

❼ Check the items that are available and ready to send.

❽ Enter remarks about the documentation that was not available to be sent. Forward the completed for to the Originator for further processing.

5.3 Originator

❾ Enter your name and date when the form is complete. Forward the completed form to the manager for final approval.

5.4 Manager

❿ If approved, sign and date then return to the Originator for filing.

SECTION 4

CUSTOMER DOCUMENTATION

Title: **CUSTOMER DOCUMENTATION CONTROL**		Number: EP-4-1 Revision: A
Prepared by:	Approved by:	

1.0 PURPOSE

The purpose of this procedure is to define the method used for distribution and review of customer furnished drawings and specifications.

2.0 APPLICABLE DOCUMENTS

Engineering Form:

E030, Customer Documentation Impact

3.0 GENERAL REQUIREMENTS

3.1 Overall clerical handling and tracking of customer incoming drawings and specifications shall be the responsibility of Document Control.

3.2 Customer drawings and specifications shall be reviewed for impact by Administration, Engineering, Manufacturing, Quality and Purchasing.

3.3 After the customer drawings and specifications are reviewed and approved they shall be forwarded to Document Control for filing.

4.0 CUSTOMER DOCUMENTATION CONTROL PROCEDURE

This procedure is divided into the following parts:

- Customer Documentation Control
- Customer Documentation Impact Form Preparation

The following procedure describes who is responsible and what they are supposed to do for each processing step.

4.1 Customer

Steps

1. Forwards drawings and specs to the company.

4.2 Mail Clerk

2. Forward customer drawings and specs to Document Control.

4.3 Document Control

3. Run copies of customer drawings and specs and retain original copy in a (pending) file. This applies to both new and revised customer drawings and specs.

4. Forward package to Engineering for impact statement.

4.4 Engineering

5. Review customer drawings and specification for impact with respect to cost, schedule, safety, etc. and forward package to Manufacturing for further processing.

4.5 Manufacturing

6. Review customer drawings and specification for impact with respect to cost, schedule, safety, etc. and forward package to Quality for further processing.

4.6 Purchasing

8. Review customer drawings and specification for impact with respect to cost, schedule, safety, etc. and forward package to Quality for further processing.

4.7 Quality

7. Review customer drawings and specification for impact with respect to cost, schedule, safety, etc. and forward package to Document Control for further processing.

4.8 Document Control

9. Forward package to Administration for further processing.

4.9 Administration

10. Summarize inputs and perform one of the following, either ***A***. or ***B***.:

    A. If there is a problem, send impact statements to customer for resolution.

11. Retain router form, drawings and Specifications awaiting Customer approval.

12. After approval, Administration forwards router form, drawings, and specifications to Document Control for logging and filing.

4.10 Document Control

13. Destroy the copies that were routed, then file the router form and the original copies that were in the pending file.

4.11 Administration

    B. If no problem, forward package to Document Control for logging and filing.

4.12 Document Control

14. Destroy the copies that were routed, then file the router form and the original copies that were in the pending file.

CUSTOMER DOCUMENTATION IMPACT			
Originator: ❶			Date:
Customer: ❷			❑ New or ❸ ❑ Revised Document
Document Number: ❹ Revision: Title:			
Reviewer ❺	Impact Yes	 No	Impact statement in regards to: Cost, Schedule, Safety, etc.
Administration	❻		❼
Engineering			
Manufacturing			
Purchasing			
Quality			
Document Control (Last) ❽			

Form E030 (Procedure EP-4-1)

Figure 35. Customer Documentation Impact Form

5.0 CUSTOMER DOCUMENTATION IMPACT FORM PREPARATION

The procedure for processing the Customer Documentation Impact form shall be followed by each individual responsible for entering information on the form. Each circled number below corresponds to the circled number on the Customer Documentation Impact form E030. (See Figure 35. for an example of the form.)

5.1 Originator

❶ Enter your Name and Date the form was originated.

❷ Enter customer name.

❸ Check appropriate box for either new documentation or revised documentation.

❹ Enter document number, revision letter and document name.

❺ Enter names of all reviewers.

5.2 All Departments

❻ Check appropiate box.

❼ Enter impact statements.

5.3 Document Control

❽ Enter impact statement, then forward the package to Administration for final summary and follow-up with the customer.

Title: **ENGINEERING CHANGE PROPOSAL**		Number: EP-4-2 Revision: A
Prepared by:	Approved by:	

1.0 PURPOSE

The purpose of this procedure is to provide instructions, and to assign responsibilities for preparing and submitting a proposed change to the customer for evaluation and disposition. It will also provide instructions for preparing the Engineering Change Proposal form.

2.0 APPLICABLE DOCUMENTS

Engineering Procedure:

EP-2-2, Document Number Assignment Logbook

Engineering Form:

E031, Engineering Change Proposal

3.0 GENERAL REQUIREMENTS

3.1 The Engineering Change Proposal is an engineering change package that consist of a completed Engineering Change Proposal form and other supplemental data, as necessary. An Engineering Change Proposal shall be provided for each proposed change which may result in a technical and/or hardware change subsequent to the establishment of an approved technical and/or hardware/software baseline.

3.2 For Engineering Change Proposal package, the following elements of data shall be included in all Engineering Change Proposals to which they apply with exception of an urgent Engineering Change Proposal. An entry shall be made in each line. "Not Applicable" (N/A) will be used only after due consideration.

3.3 The urgent Engineering Change Proposal may be utilized for all changes proposed prior to establishing the completed hardware/software baseline defined by the Configuration Baseline Document.

4.0 ENGINEERING CHANGE PROPOSAL PROCEDURE

This procedure is divided into the following parts:

- Engineering Change Proposal
- Engineering Change Proposal Form Preparation

The following procedure describes who is responsible and what they are supposed to do for each processing step.

4.1 Originator

<u>Steps</u>

1. Prepare the Engineering Change Proposal form E031 to submit a proposed change to the customer for evaluation and disposition. (See Figure 36. for an example of the form.)

2. Forward the completed Engineering Change Proposal form to your manager for review and approval.

4.2 Manager

3. Verify that the Engineering Change Proposal form is complete and correct. If approved, sign and date, then return the form to the Originator for further processing.

4.3 Originator

4. Forward the approved Engineering Change Proposal form to Document Control for the assignment of a Engineering Change Proposal number.

4.4 Document Control

5. Assign the next available number from the Document Number Assignment Logbook EP-2-2, under Paragraph 9.0, Engineering Change Proposal Number Assignment Log. Then enter the number in the Engineering Change Proposal form E031.

6. Forward a copy of the Engineering Change Proposal form to the Originator.

7. Run copies, stamp, distribute to customer, then file the original in the Engineering Change Proposal file by number.

4.5 Engineering

8. After receipt of the returned Engineering Change Proposal Request form from the customer, approved or rejected, it shall be forward to the Originator for disposition.

ENGINEERING CHANGE PROPOSAL		
Requester: ❶ Department No.: Date:		ECP No.: ❷
Manager: ❸ Department No.: Date:		Priority: ❹ ❑ Routine ❑ Urgent
Models effected: ❺		
Serial Number effectivity: ❻		
Part No.: ❼	Part Name:	Drawing No.:
Reason for the change: ❽		
Description of the change: ❾		
Approved by: (Customer) Name: ❿ Title: Date:		

Form E031 (Procedure EP-4-2)

Figure 36. Engineering Change Proposal Form

5.0 ENGINEERING CHANGE PROPOSAL FORM PREPARATION

The procedure for processing the Engineering Change Proposal form shall be followed by each individual responsible for entering information on the form. Each circled number below corresponds to the circled number on the Engineering Change Proposal form E031. (See Figure 36. for an example of the form.)

5.1 Engineer

❶ Enter your Name, Department Number and Date.

5.2 Document Control

❷ Assign the next available number from the Document Number Assignment Logbook EP-2-2, under Paragraph 9.0, Engineering Change Proposal Number Assignment Log. Then enter the number in the Engineering Change Proposal form E031.

5.3 Manager

❸ Enter your Name, Department Number and Date, upon approval.

5.4 Engineer

❹ Mark the appropriate box:

Routine - will be processed within five days.

Urgent - will be processed the same day they are received.

❺ Enter the Model number(s) that are affected by this change.

❻ Enter the Serial Number(s) that will be affected by this change.

❼ Enter the Part Number, Part Name and Drawing Number.

❽ Enter the reason for the change.

❾ Enter the description of the change.

❿ Enter your Name (Customer), Title and Date, if approved.

Title: **DEVIATION/WAIVER**		Number: EP-4-3 Revision: A
Prepared by:	Approved by:	

1.0 PURPOSE

The purpose of this procedure is to provide instructions, and to assign responsibilities for requesting, from a customer, advance permission to deviate from specified requirements or a waiver of specified requirements for an already manufactured part or assembly. It will also provide instructions for preparing and submitting the Deviation/Waiver Request form.

2.0 APPLICABLE DOCUMENTS

Engineering Procedure:

EP-2-2, Document Number Assignment Logbook

Engineering Form:

E032, Deviation/Waiver Request

3.0 DEFINITIONS

3.1 Deviation

Permission, in writing and in advance of manufacture, to deviate from specified requirements for a given number of units or for a specified period of time.

3.2 Waiver

Permission, in writing, to accept for use a completed but nonconforming item either "as-is" or upon completion of rework, this is applicable to a given number of units or for a specified period of time.

3.3 Minor

The deviation consists of a departure which does not involve any of the factors in Paragraphs: 3.4 and 3.5.

3.4 Major

The deviation consists of a departure involving (1) health, (2) performance, (3) interchangeability, reliability or maintainability of the item or its repair parts, (4) effective use or operation, (5) weight or (6) appearance (when a factor).

3.5 Critical

The deviation consists of a departure involving safety.

4.0 GENERAL REQUIREMENTS

4.1 Recurring Deviations

If a proposed deviation is recurring (a repetition or extension of a previous deviation) it is probable that either the requirements of the documentation are too stringent or the practices of the manufacturer are questionable. A determination should be made concerning a means of eliminating the need for a recurring deviation prior to its submittal and/or authorization.

5.0 DEVIATION/WAIVER PROCEDURE

This procedure is divided into the following parts:

- Deviation/Waiver Procedure
- Deviation/Waiver Request Form Preparation

The following procedure describes who is responsible and what they are supposed to do for each processing step.

5.1 Originator

Steps

1. Prepare the Deviation/Waiver Request form E032 to request, from the customer, advanced permission to deviate from specified requirements or a waiver of specified requirements for an already manufactured part or assembly. (See Figure 37. for an example of the form.)

2. Forward the completed Deviation/Waiver Request form to your manager for review and approval.

5.2 Manager

3. Verify that the Deviation/Waiver Request form is complete and correct. If approved, sign and date, then return the form to the Originator for further processing.

5.3 Originator

4. Forward the approved Deviation/Waiver Request form to Document Control for the assignment of a Deviation/Waiver Request number.

5.4 Document Control

5. Assign the next available number from the Document Number Assignment Logbook EP-2-2, under Paragraph 6.0, Deviation/Waiver Number Log. Then enter the number in the Deviation/Waiver Request form E032.

6. Forward a copy of the Deviation/Waiver Request form to the Originator.

7. Run copies, stamp, distribute to customer, then file the original in the Deviation/Waiver Request file by number.

5.5 Engineering

8. After receipt of the returned Deviation/Waiver Request form from the customer, approved or rejected, it shall be forwarded to the Originator for disposition.

<table>
<tr><th colspan="3">DEVIATION/WAIVER REQUEST</th></tr>
<tr><td colspan="2">Requester:
Department No.: ❶ Date:</td><td>Deviation No.: ❷</td></tr>
<tr><td colspan="2">Manager:
Department No.: ❸ Date:</td><td>Type of Request: ❹
❑ Deviation ❑ Waiver</td></tr>
<tr><td colspan="3">Criticality: ❑ Minor ❑ Major ❑ Critical ❺</td></tr>
<tr><td colspan="3">Program or Project: ❻ Contract No.:</td></tr>
<tr><td>Part No.: ❼</td><td>Part Name:</td><td>Drawing No.:</td></tr>
<tr><td colspan="3">Effectivity: ❽
Affects serialized assembly number: ____________________ only.
or from serialized assembly numbers: __________ through __________
or manufacturing dates: __________ through __________</td></tr>
<tr><td colspan="3">Explanation of Reason for Deviation or Waiver: ❾</td></tr>
<tr><td colspan="3">Approved by: (Customer)
Name: ❿ Title: Date:</td></tr>
</table>

Form E032 (Procedure EP-4-3)

Figure 37. Deviation/Waiver Request Form

6.0 DEVIATION/WAIVER REQUEST FORM PREPARATION

The procedure for processing the Deviation/Waiver Request form shall be followed by each individual responsible for entering information on the form. Each circled number below corresponds to the circled number on the Deviation/Waiver Request form E032. (See Figure 37. for an example of the form.)

6.1 Originator

❶ Enter your name, department number and date the form was completed.

6.2 Document Control

❷ Assign the next available part number from the Document Number Assignment Logbook EP-2-2, under Paragraph 6.0, Deviation/Waiver Request Number Assignment Log. Then enter the number in the Deviation/Waiver Request form E032.

6.3 Manager

❸ Enter your name, department number and date, if approved.

6.4 Originator

❹ Check the appropriate box depending on whether this is a request for a Deviation or a Waiver.

❺ Enter the criticality of the Deviation or Waiver.

❻ Enter the name of the program or project and its Contract Number.

❼ Enter the Part Number, Part Name, and Drawing Number.

❽ Enter the affected units, or dates of the Deviation or Waiver.

❾ Enter the explanation of the reason for the Deviation or Waiver.

6.5 Customer

❿ Enter your name (Customer), title and date, if approved.

SECTION 5

VENDOR DOCUMENTATION

Title: **VENDOR DOCUMENTATION CONTROL**		Number: EP-5-1 Revision: A
Prepared by:	Approved by:	

1.0 PURPOSE

The purpose of this procedure is to provide instructions, and to assign responsibilities for controlling company drawings and specifications that are forwarded to vendors for quotes and fabrication.

2.0 APPLICABLE DOCUMENTS

None.

3.0 DEFINITIONS

3.1 Form

The manufactured structure, shape, and material composition of an item or assembly.

3.2 Fit

The size and dimensional aspects of an item or assembly.

3.3 Function

The actual performance and level of performance of an item or assembly.

3.4 Documentation Only

Is used to indicate changes to documentation, drawings, and bills of materials, etc., that are strictly error corrections and do not effect the manufacturing process. Examples of changes include correction of spelling errors, typographical errors, non-functional drafting errors, etc.

4.0 VENDOR DOCUMENTATION CONTROL PROCEDURE

The following procedure describes who is responsible and what they are supposed to do for each processing step.

4.1 Purchasing

Steps

1. When Purchasing receives a quote from a vendor and the vendor is accepted by the company, Purchasing will prepare a Purchase Order and attach drawings and/or specifications that are at Revision level "A". This is for a new design only. When Purchasing goes out for quotes for an older design the revision level could be revision "F" or "G" etc. This procedure covers a new design revision level "A" documents because the system is the same no matter what the revision level.

2. When a change is required to a company design, Purchasing will add the following note to the Change Order to the Purchase Order. This applies every time there is a change to a drawing and/or a specification.

    DRAWING AND/OR SPECIFICATION REVISION "B" IS EFFECTIVE AS OF (date or number of units); RETURN REVISION "A" TO THE COMPANY FOR DISPOSAL.

3. Purchasing will notify the vendor that a Change Order to the Purchase Order is being forwarded to them and that they need to return revision "A" to the company as of the effective date of the new change.

4.2 Major Change (Class 1) - Form, Fit or Function

4.3 Engineering

4. If a major change is required, Engineering will notify Purchasing.

4.4 Purchasing

5. Purchasing will contact the vendor and have them stop working to the revision "A" drawings and/or specifications until further notice and that a formal change package will be forthcoming.

4.5 Engineering

6. Engineering will red-line the drawings and/or specifications and present a change package to the company Change Control Board for approval.

4.6 Change Control Board

7. The Change Control Board will decide if and when a change should be imposed on the vendor, after Change Control Board approval.

4.7 Engineering

8. Engineering will change drawings and/or specifications originals to revision "B".

4.8 Document Control

9. Document Control releases the drawings and/or specifications at revision "B". Runs copies and stamps drawings and/or specifications and forwards copies to Purchasing.

10. Following is the document stamping method used at the company to control document use at the vendor. There are three phases that vendor drawings might go through:

    a. Quote

       Stamp (red) - QUOTATION PURPOSES ONLY

    b. Development Phase

       Stamp (red) - DEVELOPMENT RELEASE

    c. Production Phase

       Stamp (red) - PRODUCTION RELEASE

4.9 Purchasing

11. Purchasing prepares a Change Order to the Purchase Order and attaches the revised drawings and/or specifications, then forwards the change package to the vendor for implementation.

12. The Change Order to the Purchase Order will state that the revision "B" to the drawings and/or specifications are effective as of a specific date or number of unit and to return revision "A" drawings and/or specifications to the company for disposal.

13. Purchasing will notify the vendor that a Change Order to the Purchase Order is being forwarded to them and that they should return revision "A" drawings and/or specifications to the company for disposal.

4.10 Minor Change (Class 2)- Documentation Only or other than Form, Fit, Function (See Major Change Paragraph 4.2).

4.11 Purchasing

14. If a minor change is required, Purchasing will notify the vendor that a minor change will be forwarded to them and that they can continue working with the revision "A" documentation. The minor change may require some rework.

4.12 Engineering

15. Engineering changes the drawings and/or specifications to revision "B".

4.13 Purchasing

16. Purchasing prepares a Change Order to the Purchase Order and attaches the revised drawings and/or specification, then forwards the change package to the vendor for incorporation.

17. The Change Order to the Purchase Order will state that revision "B" drawings and/or specifications are effective as of a specific date or number of units.

18. Purchasing calls the vendor and tells them that the Change Order to the Purchase Order is being forwarded to them and that they need to return the revision "A" drawings and/or specifications to the company for disposal.

19. The Change Order to the Purchase Order will state that the Revision "A" drawings and/or specifications will need to be returned to the company for disposal.

Title: **SINGLE SOURCE AUTHORIZATION**		Number: EP-5-2 Revision: A
Prepared by:	Approved by:	

1.0 PURPOSE

The purpose of this procedure is to provide instructions, and to assign responsibilities for obtaining approval to use purchased parts with only one procurement source. It will also provide instructions for preparing and submitting the Single Source Authorization form.

2.0 APPLICABLE DOCUMENTS

Engineering Procedure:

EP-2-2, Document Number Assignment Logbook

Engineering Form:

E033, Single Source Authorization

3.0 SINGLE SOURCE AUTHORIZATION PROCEDURE

This procedure is divided into the following parts:

- Single Source Authorization Procedure
- Single Source Authorization Form Preparation

The following procedure describes who is responsible and what they are supposed to do for each processing step.

3.1 Originator

Steps

1. Prepare the Single Source Authorization form E033 to obtain approval to use purchased parts with only one procurement source. (See Figure 38. for an example of the form.)

2. Forward the completed Single Authorization form to your manager for review and approval.

3.2 Manager

3. Verify that the Single Source Authorization form is complete and correct. If approved, sign and date, then return the form to the Originator for further processing.

3.3 Originator

4. Forward the approved Single Source Authorization form to Document Control for the assignment of a Single Source Authorization number.

3.4 Document Control

5. Assign the next available number from the Document Number Assignment Logbook EP-2-2, under Paragraph 8.0, Single Source Authorization Number Assignment Log. Then enter the number in the Single Source Authorization form E033.

6. Forward a copy of the Single Source Authorization form to the Originator.

7. Run copies, stamp, distribute to the customer, then file the original in the Single Source Authorization file by number.

SINGLE SOURCE AUTHORIZATION	
Engineer: **Department No.:** ❶ **Date:**	**Single Source Authorization No.:** ❷
Manager: **Department No.:** ❸ **Date:**	**Project/Product(s):** ❹
Manufacturer's Name: ❺ **Part Number:** **Description:**	
Reason for use: ❻	
Authorizations: **Engineering:** ❼ **Date:**	
Manufacturing: **Date:**	
Quality: **Date:**	
Purchasing: **Date:**	

Form E033 (Procedure EP-5-2)

Figure 38. Single Source Authorization Form

4.0 SINGLE SOURCE AUTHORIZATION REQUEST FORM PREPARATION

The procedure for processing the Single Source Authorization form shall be followed by each individual responsible for entering information on the form. Each circled number below corresponds to the circled number on the Single Source Authorization form E033. (See Figure 38. for an example of the form.)

4.1 Engineer

❶ Enter your name, department number and date when the form is completed.

4.2 Document Control

❷ Assign the next available part number from the Document Number Assignment Logbook EP-2-2, under Paragraph 8.0, Single Source Authorization Number Assignment Log. Then enter the number in the Single Source Authorization form E033.

4.3 Manager

❸ Enter your name, department number and date, upon approval.

4.4 Engineer

❹ Enter the Project or Product name.

❺ Enter the Manufacturer's Name, Part Number and Part Description.

❻ Enter the reason for use. Provide sufficient justification for risking production line delay or shutdown if the part cannot be obtained in the required time.

4.5 All Departments

❼ Each Department Manager, enter your name and date, upon approval.

Title: **LIMITED BUY AUTHORIZATION**		Number: EP-5-3 Revision: A
Prepared by:	Approved by:	

1.0 PURPOSE

The purpose of this procedure is to provide instructions, and to assign responsibilities for communicating situations where materials should be purchased in limited quantities or not at all. It will also provide instructions for preparing and submitting the Limited Buy Authorization form.

2.0 APPLICABLE DOCUMENTS

Engineering Procedure:

EP-2-2, Document Number Assignment Logbook

Engineering Form:

E034, Limited Buy Authorization

3.0 OVERVIEW

The Limited Buy Authorization is a tool by which Engineering may communicate situations to other affected departments within the company in which materials should be purchased only in limited quantities or not at all. Examples of such situations are Engineering plans to obsolete components but the applicable Engineering Change has not completed the approval cycle, or when a new design for release of new components or obsolescence of old components is not ready for full-scale production. Development Engineering may also use the Limited Buy Authorization during the Development Phase to communicate that limited buys are advisable.

4.0 LIMITED BUY AUTHORIZATION PROCEDURE

This procedure is divided into the following parts:

- Limited Buy Authorization Procedure
- Limited Buy Authorization Form Preparation

The following procedure describes who is responsible and what they are supposed to do for each processing step.

4.1 Originator

Steps

1. Prepare the Limited Buy Authorization form E034 to limit the number of purchased parts for an end product, basically during development. (See Figure 39. for an example of the form.)

2. Forward the completed Limited Buy Authorization form to your manager for review and approval.

4.2 Manager

3. Verify that the Limited Buy Authorization form is complete and correct. If approved, sign and date, then return the form to the Originator for further processing.

4.3 Originator

4. Forward the approved Limited Buy Authorization form to Document Control for the assignment of a Limited Buy Authorization number.

4.4 Document Control

5. Assign the next available number from the Document Number Assignment Logbook EP-2-2, under Paragraph 7.0, Limited Buy Authorization Number Assignment Log. Then enter the number in the Limited Buy Authorization form E034.

6. Forward a copy of the Limited Buy Authorization form to the Originator.

7. Run copies, stamp, distribute, then file the original in the Limited Buy Authorization file by number.

LIMITED BUY AUTHORIZATION		
Engineer: **Department No.:** ❶	**Date:**	**Limited Buy Authorization No.:** ❷
Manager: **Department No.:** ❸	**Date:**	**Models effected:** ❹
Part number and description of parts effected: ❺		
Reason for limitation: ❻		
Cost data and/or Return on Investment Analysis: ❼		
Number of days supply authorized: ❽	**Resolution Date:**	**Expiration Date:**
Authorizations: **Engineering:** ❾	**Date:**	
Manufacturing:	**Date:**	
Quality:	**Date:**	
Purchasing:	**Date:**	

Form E034 (Procedure EP-5-3)

Figure 39. Limited Buy Authorization Form

5.0 LIMITED BUY AUTHORIZATION REQUEST FORM PREPARATION

The procedure for processing the Limited Buy Authorization form shall be followed by each individual responsible for entering information on the form. Each circled number below corresponds to the circled number on the Limited Buy Authorization form E034. (See Figure 39. for an example of the form.)

5.1 Originator

❶ Enter your name, department number and date when the form is completed.

5.2 Document Control

❷ Assign the next available part number from the Document Number Assignment Logbook EP-2-2, under Paragraph 7.0, Limited Buy Authorization Number Assignment Log. Then enter the number in the Limited Buy Authorization form E034.

5.3 Manager

❸ Enter your name, department number and date, if approved.

5.4 Originator

❹ Enter the model numbers that are effected by this request.

❺ Enter the Part Number effected and its description. Only one Part Number may be entered.

❻ Enter the reason why the Limited Buy Authorization is being requested, including the impact.

❼ Enter all cost data which is available for the affected parts, including a Return on Investment Analysis.

❽ Enter Purchasing limitation in the number of days' of supply to support the Master Schedule.

Enter the resolution date.

Enter the expiration date.

5.5 All Departments

❾ Enter your name and date, upon approval.

Title: **APPROVED VENDOR LIST**		Number: EP-5-4 Revision: A
Prepared by:	Approved by:	

1.0 PURPOSE

The purpose of this procedure is to define the method used for developing and changing the Approved Vendor List.

2.0 APPLICABLE DOCUMENTS

Engineering Form:

E035, Approved Vendor List Input

3.0 DEFINITION

3.1 Approved Vendor List

A Listing of vendors qualified to supply parts and assemblies to the company for incorporation into end products. The Approved Vendor List cross references each production part number to a vendor part number and name.

4.0 RESPONSIBILITIES

4.1 Engineering is primarily responsible for the design, quality, and reliability of all parts required to manufacture products. This responsibility includes determination of qualified vendors for such parts and maintenance of the Approved Vendor List.

4.2 The Engineering Change Analyst is responsible for all data entry and inputting of changes to the Approved Vendor List database. This includes verification that all customer contractual commitments are met before any changes are input.

4.3 The Engineering Change Analyst is responsible for notifying all effected organizations within the company of any changes to the Approved Vendor List.

4.4 The Approved Vendor List Committee, composed of representatives from Engineering, Manufacturing, Purchasing and Quality, reviews any parts or vendors to be added to, changed or deleted from the Approved Vendor List. This committee serves in an advisory capacity to Engineering, which has final responsibility for the Approved Vendor List.

5.0 APPROVED VENDOR LIST PROCEDURE

This procedure is divided into the following parts:

- Approved Vendor List Procedure
- Approved Vendor List Input Form Preparation

5.1 The need for a change or an addition to the Approved Vendor List is normally determined by Engineering or Purchasing. The following are examples of the types of changes that may be made to the Approved vendor List:

A vendor may be added for additional sourcing.

A problem vendor may be deleted.

A new part number may be added per an Engineering Change that required an approved vendor.

An error, such as a typographical error in the vendor part number that may need to be corrected.

5.2 The following procedure describes who is responsible and what they are supposed to do for each processing step.

5.2.1 Originator

Steps

1. Prepare the Approved Vendor List Input form E035 to add or change vendor information in the Approved Vendor List database. (See Figure 40. for an example of the form.)

   Note: *If the originator is from a department other than Engineering, at least 10 parts samples must be attached to the Approved Vendor List Form.*

2. Forward the completed Approved Vendor List Input form to Document Control.

5.2.2 Document Control

3. Insert a copy of the Approved Vendor List Input form into the Engineering Change Package if the Approved Vendor List is part of an Engineering Change.

4. Forward the Approved Vendor List Input form and all supporting documentation, including the attached samples, to Engineering.

5.2.3 Engineering

5. Perform all necessary evaluations and testing necessary to determine the impact of the change regarding the product, and accept or reject the suggested Vendor accordingly.

6. Obtain the signature of the Manager of Engineering, if a sole source situation arises as the result of an accepted Approved Vendor.

7. If rejected, the Approved Vendor List Input form shall be sent back to Document Control, with sufficient reason and/or supporting documentation attached, where it shall be logged as rejected and returned to the originator.

8. If accepted, Engineering shall complete the Approved Vendor List Input form, and send it, with supporting documentation attached, to Document Control. Document Control shall retain the Approved Vendor List Input form on file.

9. Forward the Approved Vendor List Input form to Document Control for distribution and review prior to the Approved Vendor List Committee meeting.

5.2.4 Approved Vendor List Committee

10. The Approved Vendor List Committee shall meet to review, evaluate, and accept or reject all Approved Vendor's which Engineering has accepted.

11. If rejected, the Approved Vendor List Input form shall be returned to Document Control, with sufficient reason and/or supporting documentation attached, where it shall be logged as rejected and returned to the originator.

12. If accepted, the Approved Vendor List Input form shall be returned to Document Control, where it shall be logged in as approved. In-house, and all contractual commitments with customers shall be met at this time.

5.2.5 Document Control

13. Document Control shall forward the Approved Vendor List Input form (prepared, reviewed, and signed by Engineering), along with any supporting documentation to Marketing, retaining a copy and the original Approved Vendor List Input form.

5.2.6 Marketing

14. Forward the Approved Vendor List Input form to all customers whose contracts dictate their approval on vendor sourcing changes. The approval waiting period may vary among products, as different customer contracts may not specify the same amount of time which each customer has in which to respond.

15. Marketing shall forward the form to Document Control, who shall log the Approved Vendor List Input form in as accepted or rejected, upon receipt of the signed Approved Vendor List Input form from the customer.

5.2.7 Document Control

16. If rejected, Document Control shall forward copies of the Approved Vendor List Input form to the originator and to Engineering.

17. If accepted, Document Control shall obtain Marketing's signature on the Approved Vendor List Input form (as verification that all customer contracts have been met), and forward copies to Purchasing, Receiving Inspection, and Engineering. Document Control shall enter the new information into the database.

APPROVED VENDOR LIST INPUT	
Originator: Department No.: ❶ Date:	Associated EC No.: ❷
Model(s) Effected: ❸	Priority: ❹ ❑ Routine ❑ Urgent
Part Number: ❺ Description:	
❻ ❑ Add Vendor Name__________ Vendor Part Number:________ ❑ Change Vendor Name__________ Vendor Part Number:________ ❑ Delete Vendor Name__________ Vendor Part Number:________ ❼ Disposition (For Delete Only) Use-as-is ____ Rework ____ Scrap ____	
Reason for Change: ❽	
Justification/Impact: ❾	
Authorizations: Engineering: ❿ Date:	
Manufacturing: Date:	
Quality: Date:	
Purchasing: Date:	

Form E035 (Procedure EP-5-4)

Figure 40. Approved Vendor List Input Form

6.0 APPROVED VENDOR LIST INPUT FORM PREPARATION

The procedure for processing the Approved Vendor List Input form shall be followed by each individual responsible for entering information on the form. Each circled number below corresponds to the circled number on the Approved Vendor List Input form E035. (See Figure 40. for an example of the form.)

6.1 Originator

❶ Enter your name, department number and date the form was originated.

❷ Enter the associated Engineering Change number, if applicable.

❸ Enter all model(s) affected by the change.

❹ Check the priority of the change. (Urgent is for line down situations only.)

❺ Enter the part number and description for which the change is being requested.

❻ Check the action which is being requested for the referenced part number (all three boxes may be checked).

❼ Only used if deleted box is checked. Enter the disposition of parts being deleted.

❽ Enter the reason for change request, e.g. additional sourcing, adding per an Engineering Change, etc.

❾ Enter the justification for why the change should be made, including its impact on production or products.

6.2 All Departments

❿ Enter your name and date, if approved.

SECTION 6

DOCUMENT CHANGE CONTROL

Title: **CHANGE CONTROL SYSTEM**		Number: EP-6-1 Revision: A
Prepared by:	Approved by:	

1.0 PURPOSE

The purpose of this procedure is to outline and describe the activities and responsibilities necessary to affect a change to a product and design documentation from conception through implementation.

2.0 APPLICABLE DOCUMENTS

None.

3.0 DEFINITIONS

3.1 Engineering Change

An engineering change is an alteration in the configuration of a documented product. The alteration can be made to a part, an assembly, or the documentation controlling the product's configuration identification.

3.2 Engineering Change Package

An Engineering Change Package is comprised of documents that are written to accomplish an engineering change to a product. The Engineering Change orders a change to the documentation which controls the product's configuration.

3.2.1 Engineering Change

The Engineering Change is used to implement a change to a product. The Engineering Change also applies the following information: change criticality, description of change, reason for change, change impact, part disposition, retrofit information, and manufacturing cut-in information.

3.3 Change Control Board

The Change Control Board is a product oriented committee with the assigned responsibility of evaluating proposed functional Engineering Changes. The Change Control Board has the authority to approve or disapprove proposed changes on cost, safety, or reliability impacts. The Change Control Board will be staffed with authorized representatives from the following departments:

Engineering (Chairperson)
Manufacturing
Quality
Others, as determined by the chairperson of program guidelines

3.3.1 The chairperson of the board will be notified, prior to a meeting, when an alternate, who will have the full authority of the regularly assigned representative will attend.

3.4 Functional Change

A functional change is a mandatory or non-mandatory change to a product that effects Form, Fit, or Function, of the product. Any change that significantly impacts cost, reliability, performance, material requirements, or safety is also considered a functional change.

3.5 Mandatory Change

Mandatory changes are considered to be those which effect performance, reliability, and safety.

3.6 Non-mandatory Change

Non-mandatory changes are those which develop through cost reduction exercise or improve maintainability.

3.7 Documentation Change

Changes which correct drawing errors, improve drawing clarity, and do not effect the product or product structure, are documentation changes.

3.8 Change Class Code

The indicator of importance of an engineering change. Class codes alert the influenced organizations of the response necessary to implement an engineering change.

3.9 Impact Analysis

A study is performed when determined necessary by the Change Control Board to communicate change impact information. When completed, the study reflects the schedule and cost effects of a change.

4.0 CONCEPTION

Any employee who determines a need for a change to a product may submit a change to their manager on an approved form. Upon receipt of a suggested change, the originator's manager will evaluate the worth of the change and either approve or disapprove.

4.1 Disapproval

The proposed change will be returned to the originator with an explanation for rejection.

4.2 Approval

The change will be forwarded to the responsible engineer in the originator's organization. The responsible engineer will evaluate the change and either accept or reject it. If rejected, it will be returned it to the originator with an explanation for rejection. If approved, see Paragraph 6.0.

5.0 FORMS PREPARATION

Upon approval of the change, the responsible engineer will determine the change criticality (change class code), the disposition of material, and prepare the proper form and forward it to the Engineering Change Analyst.

Note: *A single Engineering Change under a single engineering change revision letter shall not cover unrelated engineering changes; rather, a separate Engineering Change shall be submitted for each engineering change which has its distinct objective. All Engineering Changes will be submitted to the Change Control Board for consideration.*

6.0 APPROVAL

The Engineering Change Analyst will check the proposed change for format accuracy and completeness and submit it to Engineering for preliminary review. The content or technical aspect of the change will remain the responsibility of Engineering. When satisfied, the Engineering Change Analyst will schedule the change(s) for the next Change Control Board meeting.

Note: *The Engineering Change Analyst will publish a Change Control Board minute/agenda. The minute/agenda will indicate proposed changes going to Change Control Board for initial action and changes returning to Change Control Board for additional review or final approval and the minutes of the meeting.*

If the proposed change is rejected by Engineering, it will be returned to the responsible engineer with a written explanation for rejection. If the reason for rejection is incomplete information, the responsible engineer will correct the form(s) as required and resubmit to the Engineering Change Analyst. The Change Control Board will ultimately make a decision to concur or non-concur on a proposed change as a result of the content of the form(s) submitted and the contributions of the Board members. The Change Control Board will make one of the following determinations:

6.1 Unanimous Preliminary Approval

The approved proposed change will be returned to the Engineering Change Analyst. The Engineering Change Analyst will complete all engineering forms necessary to implement the change and reschedule it to the Change Control Board. After a proposed change has had preliminary Change Control Board approval and rescheduled to the Change Control Board, it is eligible for final approval.

6.2 Unanimous Rejection

The rejected proposed change will be returned with written explanation for rejection to the Engineering Change Analyst who will return the package to the responsible engineer of the originating department.

6.3 No Decision

It may be determined by the Chairperson or other Board members that insufficient information has been made available or a high level of management approval is necessary. Action items will be assigned by the Chairperson and rescheduled to the Change Control Board.

6.4 Impact Analysis

When conducted, the impact analysis will be subjected to the following dollar value and approval authority:

Up to $ 5,000.00	Change Control Board
Over $ 5,000.00	V.P./Engineering

6.5 Final Decision

The Change Control Board will review the completed change package and make a final decision to concur or not concur. Upon concurrence, the Engineering Change Analyst will forward the total approved change package to the Document Control for distribution and filing. If a decision is made to not concur, the Engineering Change Analyst will return the package to the responsible engineer with written explanation. In either case, the final decision will be published in the minutes of the Change Control Board meeting.

7.0 DISTRIBUTION

Document Control will distribute approved copies of the change package to all holders or affected controlled documents.

4.0 IMPLEMENTATION

Drafting and the Engineering Change Analyst will be responsible for incorporating the Engineering Changes into the documentation package.

All changes to documentation incorporated at one time shall be identified by the same revision letter. The changes will be numbered sequentially to permit ready identification of a specific change. In this case, the appropriate revision letter will appear as a suffix to the part number, i.e., A, B, etc.

Upon receipt of the Engineering Change(s) and keyed by the effectivity date of the approved change, Manufacturing will update the database (routing file, process plan, etc.) and incorporate the change into the product(s).

Title: **ENGINEERING CHANGE ROUTING**		Number: EP-6-2 Revision: A
Prepared by:	Approved by:	

1.0 PURPOSE

The purpose of this procedure is to provide instructions, and to assign responsibilities for routing an Engineering Change Package through the Engineering Department.

2.0 APPLICABLE DOCUMENTS

Engineering Procedure:

EP-6-4, Engineering Change Procedure

3.0 DEFINITIONS

3.1 Engineering Review Board

A meeting that is held for the purpose of performing a review prior to the Engineering Change Pre-analysis period. All discrepancies about an Engineering Change package shall be cleared up prior to a Change Control Board Meeting.

3.2 Change Control Board

A meeting that is held for the purpose of final review and authorization of an engineering change package. This meeting is held at the conclusion of the Engineering Change pre-analysis.

4.0 ENGINEERING CHANGE ROUTING PROCEDURE

4.1 Engineering Change Analyst

The Engineering Change Analyst is responsible for the preparation, processing and routing of an Engineering Change Package. The Engineering Change Analyst shall be aware of the location of the Engineering Change Package at all times during the process of routing it through the Engineering Department.

<u>Steps</u>

1. Upon receipt of an Engineering Change Package, the Engineering Change Analyst shall examine the Package, using the Engineering Change Package Contents Sheet as a check-off list. (Reference: EP-6-4, Engineering Change Package.) The Engineering Change Analyst shall verify that all documents supplied by the originator have been filled out completely and correctly, and shall insert all documents for which they are responsible.

2. If an item is designated on the Engineering Change Package Contents Sheet and is not included in the Engineering Change Package, the Engineering Change Analyst shall contact the originator and either obtain it and insert it into the Engineering Change Package or state why it will not be included. The reason shall be written next to the item that is missing.

3. When the need for an Engineering Change is apparent, an Engineering Review Board meeting shall be scheduled by the Engineering Change Analyst. The urgency of the Engineering Change shall be taken into consideration when scheduling this meeting.

4. The Engineering Change Analyst shall make a copy of the Engineering Change Cover Sheet, stamp it with the Engineering Review Board stamp, and write in the date, time and place of the Engineering Review Board meeting.

5. The Engineering Change Package shall be taken to Document Control for Engineering Review Board distribution.

6. After the Engineering Change Package has been distributed by Document Control, the Engineering Change Package will be returned to the Engineering Change Analyst to hold until the Engineering Review Board meeting.

7. The Engineering Change Analyst will take the Engineering Change Package to the Engineering Review Board meeting. Any changes that occur during the meeting shall be marked in red. After the meeting, the Engineering Change Analyst shall take all of the red line mark-ups, separate them off with a paper clip to avoid any confusion when the Package is taken to Document Control for Change Control Board pre-analysis distribution. The paper clipped documents will eliminate the redistribution of the entire Engineering Change Package.

8. In some cases the Engineering Change Package will be signed off during the Engineering Review Board meeting. If this happens, the Engineering Change Analyst shall proceed to Step 12. and prepare the Engineering Change Package for final release distribution.

9. The Engineering Change Analyst shall schedule a Change Control Board meeting. A copy of the Engineering Change Cover Sheet shall be made and stamped with the Change Control Board meeting stamp and the date, time and place. The Engineering Change Package shall be taken to Document Control for Change Control Board pre-analysis distribution.

10. After the package has been distributed by Document Control, it will be returned to the Engineering Change Analyst to hold until the Change Control Board meeting.

11. The Engineering Change Package will be taken to the Change Control Board meeting by the Engineering Change Analyst, where it will be signed off by all Change Control Board members. All discussion of the Engineering Change should have taken place in the Engineering Review Board meeting.

12. After the Change Control Board meeting the Engineering Change Analyst will coordinate all necessary updates to all effected documents. If drawings are effected, the Engineering Change Analyst shall take the red line mark-ups with an attached copy of the Engineering Change cover sheet to Drafting so the masters can be revised. If specifications or Product Manuals are effected, the release date of the Engineering Change Package shall be written in and the Engineering Change Analyst shall take them to Technical Publications so the necessary updates can be made.

13. The Engineering Change Analyst is responsible for updating all effected areas on the Computer system, including the Item Master, Bill of Material and Approved Vendor List. Updating shall be accomplished promptly so that changes can be implemented throughout the company. This shall be done while Drafting and Technical Publications are updating the documents for which they are responsible.

14. The Engineering Change Analyst is responsible for assuring that all required updates have been made. The Engineering Change Analyst is also responsible for obtaining all approval signatures on all applicable updated documents excluding the applicable drawing signatures, the Draftsperson shall obtain these, before giving the Engineering Change Package to Document Control.

15. Upon completion of all the updates, the Engineering Change Package shall be taken to Document Control for final release distribution and filing.

4.2 Document Control

Document Control is responsible for making copies of Engineering Change Packages for the Engineering Review Board, Change Control Board pre-analysis, and final released distribution. Other responsibilities include microfilming and filing of Engineering Change packages, the individual filing of the specifications and master drawings and maintaining Engineering Change Package distribution lists.

1. Upon receipt of an Engineering Change Package for Engineering Review Board distribution Document Control shall verify that there is a copy of the Engineering Change Cover Sheet that has been stamped with the Engineering Review Board stamp and date, time and place. The Engineering Change Package shall be returned to the Engineering Change Analyst for correction if this has not been completed.

2. Using the Engineering Change Package Preparation Guide as a check-off list, Document Control shall make copies of every document that is listed for pre-analysis distribution. It is the responsibility of the Document Control to assure that the Engineering Change Package is distributed prior to the scheduled Engineering Review Board meeting.

3. The Engineering Change Package shall be returned to the Engineering Change Analyst to hold until the Engineering Review Board meeting.

4. After the Engineering Review Board meeting, the Engineering Change Package shall be returned to Document Control for Change Control Board pre-analysis distribution. Document Control shall verify that there is a copy of the Engineering Cover Sheet that has been stamped with the Change Control Board stamp and the date, time and place. The Engineering Change Package shall be returned to the Engineering Change Analyst for correction if this has been completed.

5. The changes that occurred during the Engineering Review Board meeting shall be paper clipped together by the Engineering Change Analyst and will be the only things besides the Engineering Change Cover Sheet that will be distributed for review prior to the scheduled Change Control Board meeting. Distribution shall be done promptly to allow 3-5 days (depending on the urgency of the Engineering Change) review time.

6. When distribution has been completed, Document Control shall return the Engineering Change Package to the Engineering Change Analyst to hold until the Change Control Board meeting.

7. After the Engineering Change Package has been completely signed off by all Change Control Board members, Document Control will receive an Document Original Withdrawal form from Drafting requesting the masters of the affected drawings. It is the responsibility of Document Control to promptly pull all drawing masters to fill these requests.

8. The Document Control Clerk will sign the Document Original Withdrawal form upon completion of pulling the document master(s). The Document Master Withdrawal form shall be filed in Document Control until all of the originals are returned.

9. Upon completion of all updates to the Engineering Change Package the Engineering Change Package shall be returned to Document Control for final distribution and Microfilming. Document Control shall verify that all Document originals have been returned; if so, the designated column on the Withdrawal form shall be filled in and the form shall be filed for future reference.

10. Using the Engineering Change Package Preparation Guide as a check-off list, Document Control shall make copies of every document that is listed for final distribution and distribute them.

11. Upon completion of final distribution, Document Control shall prepare the Engineering Change Package to be microfilmed. Using the Engineering Change Preparation Guide as a check-off list, Document Control shall remove the designated documents that will be microfilmed and microfilm them.

12. After microfilming is complete, Document Control shall file the master drawings, specifications and affected pages of the Product manuals separately from the Engineering Change Package in their designated places for future reference.

4.3 Drafting

The Drafting department is responsible for updating all Engineering Change affected original drawings.

1. Upon receipt of red lined mark-ups from the Engineering Change Analyst, Drafting shall proceed to fill out an Document Original Withdrawal form requesting the affected drawing master(s) from Document Control.

2. After the request has been filled by Document Control, Drafting shall take the master drawing(s) and make all necessary updates. Updating shall be done promptly (depending on the urgency of the Engineering Change) so that changes can be implemented throughout all departments that are affected.

3. It shall be the responsibility of Drafting to obtain all applicable drawing approval signatures. The updated master drawings and their red lined mark-ups shall be returned to the Engineering Change Analyst upon completion of the above tasks.

4.4 Technical Publications

The Technical Publications Department is responsible for reviewing and updating all specifications and Product Support manuals.

1. It is the responsibility of the Technical Publications writer to review all Engineering Change Packages and to assure that if Product manuals are effected, the red line mark-ups are included in the Engineering Change Package. Technical Publications shall also assure that the Engineering Change Cover Sheet states that Product manuals are effected; if not, the Technical Publications writer shall contact the originator of the Engineering Change Package prior to the Change Control Board meeting.

2. Technical Publications is also responsible for editing any specifications during pre-analysis to assure their correctness and to assure that they conform to the established standards and formats. Discrepancies shall be cleared up with the originator of the Engineering Change Package.

3. If necessary, the Technical Publications writer shall attend the Engineering Review Board meeting to discuss and clear up all discrepancies regarding specifications and Product, manuals.

4. Upon receipt of a red line marked-up specification the Technical Publications writer shall create a new updated original. When completed, the red lined specification and the new updated original shall be returned to the Engineering Change Analyst.

5. When a red lined Product Support manual is received, the Technical Publications writer shall generate a new printout.

6. The Technical Publications writer shall take the printout of the manual to a Technical Publication Illustrator. The Illustrator shall generate a new master and send it out for printing. When completed, the affected page(s) of the Product manual and the new updated original(s) shall be returned to the Engineering Change Analyst. If the Product Manual is being initially released the Engineering Change Analyst will receive the entire original Product Manual.

Title: **ENGINEERING CHANGE REQUEST**		Number: EP-6-3 Revision: A
Prepared by:	Approved by:	

1.0 PURPOSE

The purpose of this procedure is to define the steps required to request changes to product documentation that will require some in-depth analysis and how to process the Request for Analysis form along with its associated responsibilities. The Engineering Procedure is used for changes that do not need an analysis.

2.0 APPLICABLE DOCUMENTS

Engineering Procedure:

EP-2-2, Document Number Assignment Logbook

Engineering Form:

E036, Request For Analysis

3.0 GENERAL REQUIREMENTS

3.1 The Request For Analysis is a formal document initiated by any person who has identified a problem with a product, or has a cost saving or product improvement idea. Requests for changes to the design or product documentation of products shall be directed to Engineering on a Request For Analysis form.

3.2 New design documentation that has never been released and documentation only changes shall be entered into the documentation system with an Engineering Change, therefore, bypassing this Request for Analysis procedure.

3.3 The Request For Analysis may be required as a follow-up to a deviation if the temporary change needs to be made permanent.

3.4 Each Request For Analysis shall request a single change which may effect more than one part, assembly, drawing, and/or other document. A Request For Analysis shall be limited to only one problem.

4.0 ENGINEERING CHANGE REQUEST PROCEDURE

This procedure is divided into the following parts:

- Engineering Change Request Procedure
- Request For Analysis Form Preparation

The following procedure describes who is responsible and what they are supposed to do for each processing step.

4.1 Originator

Steps

1. Prepare the Request For Analysis form E036 to request an analysis of a problem with a product or its documentation. (See Figure 41. for an example of the form.)

2. Forward the completed Request For Analysis form to your manager for review and approval.

4.2 Manager

3. Verify that the Request For Analysis form is complete and correct. If approved, sign and date, then return the form to the Originator for further processing.

4.3 Originator

4. Forward the approved Request For Analysis form to Document Control for the assignment of a Request For Analysis number.

4.4 Document Control

5. Assign the next available number from the Document Number Assignment Logbook EP-2-2, under Paragraph 16.0, Request For Analysis Number Assignment Log. Then enter the number in the Request For Analysis form E036.

6. Forward a copy of the Request For Analysis form to the Originator for further processing.

REQUEST FOR ANALYSIS					
Originator: Department No.: ❶ Date:			RAF No.: ❷		
Manager: Department No.: ❸ Date:			Model(s) Effected: ❹		
Doc. No.s	Revision	Description	Class	Type	Sheet
❺					

Description of Analysis Requested: (Problem, possible Solutions, Indication of Quality or Cost Improvements, etc.)
❻

Action To Be Taken	
❑ Urgent ❑ Routine ❑ Assign as Engineering Change No. ________ ❼ Engineering: Date:	❑ Assign as Engineering Project No. ________ Design Engineering: Date:
❑ No Action Taken Due to To: ❽ By: Date:	

Form E036 (Procedure EP-6-3)

Figure 41. Request For Analysis Form

5.0 REQUEST FOR ANALYSIS FORM PREPARATION

The procedure for processing the Request For Analysis form shall be followed by each individual responsible for entering information on the form. Each circled number below corresponds to the circled number on the Request For Analysis form E036. (See Figure 41. for an example of the form.)

5.1 Originator

❶ Enter your Name, Department Number and the Date.

5.2 Document Control

❷ Assign the next available part number from the Document Number Assignment Logbook EP-2-2, under Paragraph 16.0, Request For Analysis Number Assignment Log. Then enter the number in the Request For Analysis form E036.

5.3 Manager

❸ Enter your Name, Department Number and Date, upon approval.

5.4 Originator

❹ Enter the Model(s) that may be effected by this change.

❺ Enter the Document Numbers, Current Revisions, Descriptions;

Class, enter one of the following: "N" for Non-Interchangeable, "I" for Interchangeable, "R" New Release, or "D" for Documentation Only.

Type, enter the type of document (e.g. drawing, Bill of Material, Where Used, Specification, etc.) and the Sheet number of sheets.

❻ Enter the description of the problem, cost savings, or product improvement as clearly as possible.

5.5 Engineer

❼ Check the appropriate box to assign the request to either routine or urgent. Enter the Engineering Change number assigned to this problem. Sign and date the form if it is assigned as an Engineering Change.

or, assign as an Engineering Project. Enter number that was assigned to the project. Sign and date the form if it is assigned as an Engineering Project.

❽ Give a complete description of way there was no action taken. Sign and date form when it is complete.

Title: **ENGINEERING CHANGE PROCEDURE**		Number: EP-6-4 Revision: A
Prepared by:	Approved by:	

1.0 PURPOSE

The purpose of this procedure is to define the process, responsibilities, and requirements for the generation and release of engineering changes.

2.0 APPLICABLE DOCUMENTS

Engineering Form:

E037, Engineering Change

3.0 OVERVIEW

Changes to products shall be documented on an Engineering Change form. Each Engineering Change shall request a single change that may effect more than one part, assembly, drawing, or other document. An Engineering Change shall be limited to one problem only.

New design documentation shall be entered into the documentation system with an Engineering Change form and a Document Release Notice form.

4.0 DEFINITIONS

4.1 Interchangeable

A part that (1) possesses such physical characteristics as to be equivalent in reliability, and maintainability, to another part; and (2) is capable of being exchanged for the other item (a) without alteration for fit, and (b) without alteration of adjoining items.

4.2 Non-Interchangeable

A change that is not interchangeable that effects form, fit, and function.

Form -The manufactured structure, shape, and material composition of an item or assembly.

Fit - The size and dimensional aspects of an item or assembly.

Function - The actual performance and level of performance of an item or assembly.

4.3 Documentation Only

Documentation Only is used to indicate changes to documentation, drawings, and bills of materials, etc., that are strictly error corrections and do not effect the manufacturing process. Examples of these changes include, correction of spelling errors, typographical errors, non-functional drafting errors, etc.

This disposition is not to be used to correct or change quantities, tolerances, dimensions, or support documentation that is changing to reflect a hardware change.

5.0 ENGINEERING CHANGE PROCEDURE

This procedure is divided into the following parts:

- Engineering Change
- Engineering Change Form Preparation

The following procedure describes who is responsible and what they are supposed to do for each processing step.

5.1 Originator

Steps

1. Prepare the Engineering Change form E037 to introduce a change into the system. (See Figure 42. for an example of the form.)
2. Forward the completed Engineering Change form to your manager for review and approval.

5.2 Manager

3. Verify that the Engineering Change form is complete and correct. If approved, sign and date, then return the form to the Originator for further processing.

5.3 Originator

4. Forward the approved Engineering Change form to Document Control for the assignment of an Engineering Change number.

5.4 Document Control

5. Assign the next available number from the Document Number Assignment Logbook EP-2-2, under Paragraph 17.0, Engineering Change Number Assignment Log. Then enter the number in the Engineering Change form E037.

6. Forward a copy of the Request For Analysis form to the Originator for further processing.

<table>
<tr><th colspan="2">ENGINEERING CHANGE</th></tr>
<tr><td>Originator:
Department No.: ❶ Date:</td><td>Eng. Change No.:
❷</td></tr>
<tr><td>Manager:
Department No.: ❸ Date:</td><td>Priority: ❹
❑ Routine ❑ Urgent</td></tr>
<tr><td colspan="2">Class of Change: ❺
1) Non-Interchangeable 2) Interchangeable 3) New Document 4) Documentation Only</td></tr>
<tr><td colspan="2">Reason for Change: ❻ Check all that apply:
❑ New Release ❑ Facilitate Production ❑ Customer Directed
❑ Design Error ❑ Facilitate Inspection ❑ Vendor Requested
❑ Drafting Error ❑ Schedule Improvement ❑ Safety Improvement
❑ Redraw Error ❑ Cost Improvement ❑ Other Explain</td></tr>
<tr><td colspan="2">Affected Documents: ❼
Document Number: Revision: Title:</td></tr>
<tr><td colspan="2"></td></tr>
<tr><td colspan="2"></td></tr>
<tr><td colspan="2"></td></tr>
<tr><td colspan="2"></td></tr>
<tr><td colspan="2">Description of Change: (Or see attached marked-up documents.): ❽</td></tr>
<tr><td colspan="2"></td></tr>
<tr><td colspan="2"></td></tr>
<tr><td colspan="2"></td></tr>
<tr><td colspan="2">Authorizations:
Engineering: ❾ Date:</td></tr>
<tr><td colspan="2">Manufacturing: Date:</td></tr>
<tr><td colspan="2">Quality: Date:</td></tr>
<tr><td colspan="2">Purchasing: Date:</td></tr>
</table>

Form E037 (Procedure EP-6-4)

Figure 42. Engineering Change Form

6.0 ENGINEERING CHANGE FORM PREPARATION

The procedure for processing the Engineering Change form shall be followed by each individual responsible for entering information on the form. Each circled number below corresponds to the circled number on the Engineering Change form E037. (See Figure 42. for an example of the form.)

6.1 Non-Interchangeable and Interchangeable Changes

6.1.1 Originator

❶ Enter your Name, Department Number and the Date when the Engineering Change is completed.

6.1.2 Document Control

❷ Assign the next available part number from the Document Number Assignment Logbook EP-2-2, under Paragraph 17.0, Engineering Change Number Assignment Log. Then enter the number in the Engineering Change form E037.

6.1.3 Manager

❸ Enter your name and date, upon approval.

6.1.4 Originator

❹ Check the appropriate box either routine or urgent.

❺ For Interchangeable and Non-Interchangeable changes circle either 1 or 2;

For New Release changes circle 3;

For Documentation Only changes circle 4.

❻ Check the appropriate box(s) that best describes the reason for this change.

❼ Listed effected documents, document number, revision and title.

❽ Enter the description of the change.

6.1.5 All Departments

❾ Enter the signature of the responsible Engineer, Manufacturing, Purchasing, and Quality personnel and the date the document was approved.

Title: **DOCUMENT MARK-UP FOR ENGINEERING CHANGE PACKAGES**		Number: EP-6-5 Revision: A
Prepared by:	Approved by:	

1.0 PURPOSE

The purpose of this procedure is to provide instructions, and to assign responsibilities for red lining documents that will become part of the Engineering Change Package.

2.0 APPLICABLE DOCUMENTS

None.

3.0 DEFINITIONS

3.1 Mark-up

This term is used to refer to the drawing or document on which a change has been marked-up.

3.2 Original

The mark-up is considered to be the working master until the original drawing or document has been updated.

3.3 Exaggeration Views

When features are too small to be drawn clearly, they must be drawn to a larger scale than is shown in the principal view and in greater detail to legibly identify the change.

3.4 Spacing and Identification

Ample space must be used between views to permit placing of dimensions without crowding and to preclude the possibility of pertaining to one view overlapping or crowding the other views. When standard orthographic projections are used, it is necessary to identify or name the views.

4.0 DOCUMENT MARK-UP FOR ENGINEERING CHANGE PACKAGES PROCEDURE

4.1 Necessary Inclusions

A mark-up needs to show only the information necessary to clearly delineate a change. Excessive illustration of superfluous detail must be avoided, but never at the expense of mark-up clarity or completeness. For example: where a feature of an item occurs many times in a continuous and regular pattern, only sufficient information to illustrate the feature and its regularity of repetition is necessary, e.g., rows of bolts or rivets, features in a bolt circle, repetition of slots, splines or gears, or any other details of a repetitious nature.

4.2 Changed Data

The old or down-level data which is to be changed must only be cross-hatched (and not completely obliterated) such that the old data can be read.

4.3 Lettering

Lettering must be freehand, vertical, and upper case. The size of lettering and spacing between lettering and lines of lettering is determined by the size of the original mark-up. Minimum lettering height is 1/8 of an inch, and larger characters should be used when legibility so requires.

4.4 Mark-ups

A red ink pen must be used to identify changes which are to be implemented. Changes must be highlighted on the document, and legibility must be maintained. Careful freehand sketching is permitted as long as reproducibility is maintained.

4.5 Engineering Change History

The original copy must include all previous changes made to the part effected, as indicated by the latest released engineering change level.

Title: **ENGINEERING CHANGE COSTING**	Number: EP-6-6 Revision: A
Prepared by:	Approved by:

1.0 PURPOSE

The purpose of this procedure is to provide instructions, and to assign responsibilities for determining the cost of an engineering change. It will also provide instructions for preparing and submitting the Cost Analysis Report form.

2.0 APPLICABLE DOCUMENTS

Engineering Procedure:

EP-2-2, Document Number Assignment Logbook

Engineering Form:

E038, Cost Analysis Report

3.0 GENERAL REQUIREMENTS

3.1 All Cost Analysis Reports shall be forwarded by Change Control Board members to Document Control. Document Control will use the form for final Change Control Board Processing.

3.2 Each Cost Analysis Report shall become a part of its associated Engineering Change, and shall be numbered and identified as such.

3.3 A Cost Analysis Report is required if the Engineering Change calls for labor or material changes. The associated Engineering Change cannot be incorporated until the approved report is included.

3.4 Each part added, deleted, and/or changed per an Engineering Change must be documented on the Cost Analysis Report and counted as a part of the total system cost change (plus or minus).

3.5 There shall be no areas left blank on the Cost Analysis Report form. If any area on the form is not effected by the Engineering Change, the person(s) responsible for the Cost Analysis report completion shall enter not applicable (N/A) in the appropriate space(s).

4.0 ENGINEERING CHANGE COSTING PROCEDURE

This procedure is divided into the following parts:

- Engineering Change Costing
- Cost Analysis Report Form Preparation

The following procedure describes who is responsible and what they are supposed to do for each processing step.

4.1 Originator

Steps

1. Prepare the Cost Analysis Report form E038 to determine the cost associated with making an engineering change. (See Figures 43. and 44. for an example of the form.)

2. Forward the completed Cost Analysis Report form to your manager for review and approval.

4.2 Manager

3. Verify that the Cost Analysis Report form is complete and correct. If approved, sign and date, then return the form to the Originator for further processing.

4.3 Originator

4. Forward the approved Cost Analysis Report form to Document Control for the assignment of a Cost Analysis Report number.

4.4 Document Control

5. Assign the next available number from the Document Number Assignment Logbook EP-2-2, under Paragraph 10.0, Cost Analysis Report Number Assignment Log. Then enter the number in the Cost Analysis Report form E038.

6. Forward a copy of the Cost Analysis Report form to the Originator.

7. Run copies, stamp, distribute to the Change Control Board members, then file the original in the Cost Analysis Report file by number.

COST ANALYSIS REPORT

Engineer: ❶
Department No.: Date:

Cost Analysis Report No.: ❷

Manager: ❸
Department No.: Date:

Part Number: ❹

Application Code: ❺

❑ Added New Part ❑ Added Existing Part ❑ Change Quantity

❑ Deleted Part ❑ Fabricated Part Changed

Materials: ❻

Unit Cost Change...Was $______ Is $______ = $ ______

Quantity Used Per One System x ______

Factory Labor Cost Change +$ ______

Total Single System Cost Change + or - = $ ______

Estimated Monthly Usage (Quantity Per system X Build Schedule) ______

Standard Lead Time ______

Rework Lead Time Vendor ______

Field: ❼

Field effected ❑ Yes ❑ No Number Installed ____ Quantity Required ____

Quantity Required Spares ____ Quantity Required Systems Shipped ____

Field Logistics Affected ❑ Yes ❑ No $ ______

Field Documentation Affected ❑ Yes ❑ No +$ ______

Total Field Costs = $ ______

Form E038 (Procedure EP-6-6) Page: 1 of 2

Figure 43. Cost Analysis Report Form (Page 1 of 2.)

COST ANALYSIS REPORT

Material Handling: ❽

Tooling or Premium Charges $ __________

Scrap Charges +$ __________

Restock Charges +$ __________

Purchase Order Change/Cancel Charges +$ __________

Total Material Handling Charges = $ __________

Rework Costs: ❾

Work In Progress Affected ❑ Yes ❑ No W.I.P Quantity ____

Unit Rework Cost (In-House) Labor $_______ + Material $_______ = $__________

Unit Rework Cost (Vendor) Labor $_______ + Material $_______ = $__________

Finished Goods Affected ❑ Yes ❑ No Finished Goods Quantity ____

Unit Rework Cost (Finished Goods) Labor $________ + Materials $_______ = $__________

Total W.I.P Rework Cost (W.I.P Unit Cost x Quantity) $__________

Total Finished Goods Rework Cost (Finished Goods Unit Cost x Quantity) +$ __________

Total Rework Cost = $__________

Authorizations: ❿

Engineering: Date:

Manufacturing: Date:

Quality: Date:

Purchasing: Date:

Form E038 (Procedure EP-6-6) Page: 2 of 2

Figure 44. Cost Analysis Report Form (Page 2 of 2.)

5.0 COST ANALYSIS REPORT FORM PREPARATION

The procedure for processing the Cost Analysis Report form shall be followed by each individual responsible for entering information on the form. Each circled number below corresponds to the circled number on the Cost Analysis Report form E038. (See Figures 43. and 44. for an example of the form.)

5.1 Engineer

❶ Enter your Name, Department Number and the Date.

5.2 Document Control

❷ Assign the next available part number from the Document Number Assignment Logbook EP-2-2, under Paragraph 10.0, Cost Analysis Report Number Assignment Log. Then enter the number in the Cost Analysis Report form E038.

5.3 Manager

❸ Enter your Name, Department Number and Date, upon approval.

5.4 Engineer

❹ Enter part number of the effected part.

❺ Mark the appropriate box:

Added New Part

Added Existing Part

Change Quantity

Deleted Part

Fabricated Part Changed

❻ Enter appropriate Materials information.

Unit Cost Change Was/Is - Enter the average cost to the company to obtain one part. If applicable, in the designated spaces, enter the price as it is before and after the change. Enter the difference on the line provided.

Quantity Used Per One System - Enter the quantity of the part per system required to completely assemble one system.

Factory Labor Cost Change - Enter the + or - difference in factory labor costs which would be incurred as a result of implementing the change.

Total Single System Cost change + or - The total system cost change is determined as follows:

	Quantity per System
X	Unit Cost Change
+	Factory Labor Cost Change
=	Total System Cost Change

Estimated monthly Usage - Enter the estimated usage of the part, to be determined from the quantity used per system multiplied by the current production build schedule.

Standard lead Time - Enter the amount of time required to obtain the part from the appropriate source.

Rework Lead Time (Vendor) - Enter the amount of time required to obtain reworked parts from the appropriate source.

❼ Enter the appropriate Field information.

Field Affected Yes/No - Check appropriate box and enter the number of systems that are effected and the number of parts that will be required to maintain Field installations.

Quantity Required Spares - Enter the quantity of parts required to up date spares and the quantity to update systems which have been shipped.

Field Logistics Affected - Check appropriate box.

Field Documentation Affected - Check appropriate box.

Total Cost for Field Units - Enter the cost for implementing the change at customer locations.

❽ Enter appropriate Material Handling information.

Tooling or Premium Charges - Enter any set-up charges, and/or other tooling costs attributed to the affected part.

Scrap Charges - Enter the cost incurred as a result of scrapping any parts which are in inventory or in process at a vendor.

Restock Charges - Enter the cost to be incurred as a result of returning parts to the source vendor.

Purchase Order Change/Cancel Charges - Enter the charge of the vendor to change material on order, or to accept returned parts.

Total Material Handling Charges - Determine as follows:

	Tooling or Premium Charges
+	Scrap Charges
+	Restock Charges
+	Purchase Order Change/Cancel Charges
=	Total Material Handling Charges

❾ Enter appropriate Rework Costs information.

Work in Progress Effected Yes/No - Check appropriate box and enter the number of parts required to sustain work in progress.

Unit Rework Costs (In-House) Labor/Material - Enter the In House costs for both labor and material required to incorporate the change on one work in progress system.

Unit Rework Costs (Vendor) Labor/Material - Enter the appropriate vendor costs for both labor and material required to incorporate the change on one work in progress system.

Finished Goods Affected Yes/No - Check appropriate box and enter the total number of subject parts required to implement the change in Finished Goods.

Unit Rework Cost (Finished Goods) Labor/Material - Enter the cost of both labor and material required to implement the change in Finished Goods.

Total Work in Progress Rework Cost - Enters the total cost by multiplying the W.I.P. quantity.

Total Finished Goods Rework Cost - Enter the total cost by multiplying the Finished Goods Unit Cost times the Finished Goods quantity.

Total Rework Cost - The Total Rework Cost is determined by adding the Total W.I.P. Rework Cost to the Total Finished Goods Rework Cost.

⑩ Each Department representative enters their Name and Date, if approved.

Title: **PROTOTYPE REPORT**		Number: EP-6-7 Revision: A
Prepared by:	Approved by:	

1.0 PURPOSE

The purpose of this procedure is to provide instructions, and to assign responsibilities for determining if an engineering change functions properly prior to production. It will also provide instructions for preparing and submitting the Prototype Report form.

2.0 APPLICABLE DOCUMENTS

Engineering Procedure:

EP-2-2, Document Number Assignment Logbook

EP-6-3, Engineering Change Request

Engineering Form:

E039, Prototype Report

3.0 DEFINITION

When a change in a design effects form, fit or function of a part or assembly, or when an initial released product requires testing before it is known to function properly, the Prototype Report becomes that part of an Engineering Change Request which records a test of the change or development, insuring its reliability and manufacturability.

4.0 GENERAL

4.1 All Prototype Reports shall become part of its associated Engineering Change Request, and shall be numbered and identified as such. (Reference: EP-6-3, Engineering Change Request.)

4.2 If a Prototype Report is required, the associated Engineering Change cannot be released until the approved report is included.

4.3 When a Prototype Report is included in an Engineering Change Request, incorporation of the change and/or production of the initial release product must be completely compatible with the prototype representation both functionally and physically.

5.0 PROTOTYPE REPORT PROCEDURE

This procedure is divided into the following parts:

- Prototype Report
- Prototype Report Form Preparation

The following procedure describes who is responsible and what they are supposed to do for each processing step.

5.1 Engineer

Steps

1. Prepare the Prototype Report form E039 to determine the feasibility of making an engineering change. (See Figure 45. for an example of the form.)

2. Forward the completed Prototype Report form to your manager for review and approval.

5.2 Manager

3. Verify that the Prototype Report form is complete and correct. If approved, sign and date, then return the form to the Originator for further processing.

5.3 Engineer

4. Forward the approved Prototype Report form to Document Control for the assignment of a Prototype Report number.

5.4 Document Control

5. Assign the next available number from the Document Number Assignment Logbook EP-2-2, under Paragraph 11.0, Prototype Report Number Assignment Log. Then enter the number in the Prototype Report form E039.

6. Forward a copy of the Prototype Report form to the Originator.

5.5 Engineer

7. Order all parts necessary to assemble the desired number of prototypes of the change.

8. Assembles the prototypes and submit them to the testing department.

PROTOTYPE REPORT	
Engineer: Department No.: ❶ Date:	Prototype Report No.: ❷
Manager: Department No.: ❸ Date:	❑ New Product ❹ ❑ Existing Product
Description of Prototype: ❺	
Recommendation: ❻ ❑ Revise & Reprototype ❑ Revise & Release ❑ Release as is.	
Part No.: ❼ Part Name: Drawing No.:	
Conditions of Test: ❽	
Results: ❾	
Authorizations: Engineering: ❿ Date:	
Manufacturing: Date:	
Quality: Date:	
Purchasing: Date:	

Form E039 (Procedure EP-6-7)

Figure 45. Prototype Report Form

6.0 PROTOTYPE REPORT FORM PREPARATION

The procedure for processing the Prototype Report form shall be followed by each individual responsible for entering information on the form. Each circled number below corresponds to the circled number on the Prototype Report form E039. (See Figure 45. for an example of the form.)

6.1 Engineer

❶ Enter your Name, Department Number and the Date.

6.3 Manager

❷ Enter your Name, Department Number and Date, upon approval.

6.2 Document Control

❸ Assign the next available part number from the Document Number Assignment Logbook EP-2-2, under Paragraph 11.0, Prototype Report Number Assignment Log. Then enter the number in the Prototype Report form E039.

6.4 Engineer

❹ Mark the appropriate box to indicate the type of prototype: a change to a new product, or a change to an existing product.

❺ Enter the description of the prototype.

❻ Mark the appropriate box for the recommended action to be taken.

❼ Enter the prototype Part number, Name, and Drawing Number if one exists.

❽ Outline the set-up used to test the prototype, including parts added and\or deleted. The Conditions of the test should be outlined in as much detail as possible.

❾ Describe in detail all determinations made during the test period, including all corrections required to finalize the change for release.

❿ Each Department representative enters their Name and Date, if approved.

Title: **CHANGE CONTROL BOARD**		Number: EP-6-8 Revision: A
Prepared by:	Approved by:	

1.0 PURPOSE

The purpose of this procedure is to establish the function, authority, and responsibility of the Change Control Board. It defines the Change Control Board and explains the schedule, function, and responsibility of both the committee as a whole and of each committee member.

2.0 APPLICABLE DOCUMENTS

None.

3.0 DEFINITION

The Change Control Board is a committee whose function it is to review and evaluate all proposed changes to products. This committee is comprised of a representative from each major functional area of the Company.

4.0 MAJOR FUNCTIONAL AREAS

Engineering
Engineering Change Analyst (Chairperson)
Manufacturing
Purchasing
Quality

5.0 SCHEDULE

5.1 The Change Control Board meets once a week at a location predesignated by the Engineering Change Analyst.

5.1.1 It is imperative that all Change Control Board representatives be punctual. If one or more members are not present at the specified time, the meeting shall be cancelled.

5.1.2 The meeting shall be chaired by the Engineering Change Analyst who shall determine the meeting agenda.

5.1.3 Each Engineering Change shall be reviewed, one at a time, by the entire committee.

5.1.4 The Engineering Change shall be discussed as a single package. If there is a question with any other Engineering Change, it shall be discussed by the committee as a separate item.

5.2 Each member of the Change Control Board shall receive pre-analysis copies of each Engineering Change package to be reviewed the following week. Any useful reference material associated with the Engineering Change shall be provided by the Engineering Change Analyst and included in this distribution.

5.2.1 In order to provide each member ample time to examine the proposed changes, these copies shall be distributed by the Engineering Change Analyst at least five work days prior to the meeting.

5.2.2 Each member shall review each Engineering Change package thoroughly during the pre-analysis period.

5.2.3 Should any Change Control Board member question a proposed change, they shall contact the Engineering Change Analyst or the Responsible Engineer during the pre-analysis period, prior to the Change Control Board meeting, for clarification.

6.0 AUTHORITY/RESPONSIBILITY

6.1 Engineering, Manufacturing, Purchasing and Quality shall be represented at the weekly meeting by at least one assigned Change Control Board member with signature authority.

6.2 Should a Change Control Board member from any of the departments listed above be unable to attend the meeting, they shall send an appointed alternate who may sign in that member's place.

6.3 In order for an Engineering Change to be approved, signature from members of Engineering, Manufacturing, Purchasing and Quality are required. Signatures from other members shall be required only if the change directly affects their individual departments. No Engineering Change shall be incorporated until it contains all required approval signatures.

6.3.1 Attendance at the weekly meeting by a representative from each of the four major functional areas listed above, or by their alternate, is mandatory. If one or more of these representatives are not present, the meeting shall be cancelled.

6.3.2 Attendance at the weekly meeting is not required of any members who are not representing the four major functional areas listed above. If a member is not present, their approval signature is automatically implied, and shall be designated as such by the Change Control Board Chairperson.

6.4 After a general consensus has been reached for each Engineering Change, only the Chairperson's signature shall be required on the form.

6.5 Cost Analysis Report - During the pre-analysis period, the Cost Analysis Report shall be investigated and completed by appointed members of the Change Control Board. Completed copies shall be provided to the Engineering Change Analyst at least one (1) day prior to the Change Control Board meeting.

6.6 Effectivity dates shall be determined by a Change Control Board representative from the Materials Department, who shall confer with members of Field Engineering and Purchasing for realistic deadlines. This decision shall be made prior to the Change Control Board meeting, and reviewed for concurrence by Board members at the meeting.

6.6.1 Each Change Control Board member is responsible to see that all approved changes are implemented within his/her functional area within the agreed-to effectivity time frame.

Title: **ENGINEERING CHANGE IMPLEMENTATION**		Number: EP-6-9 Revision: A
Prepared by:	Approved by:	

1.0 PURPOSE

The purpose of this procedure is to define the requirements for determining the effectivity date and for implementing an engineering change into the manufacturing process.

2.0 APPLICABLE DOCUMENTS

None.

3.0 RESPONSIBILITY

3.1 Manufacturing has the responsibility for organizing and managing the implementation process and for tracking the progress of an Engineering Change implementation.

3.2 The Manufacturing, Production Control, Purchasing, Quality, Accounting, Sales/Marketing and Engineering departments have specific responsibilities as detailed in this procedure. Each department is responsible for reviewing proposed Engineering Changes and for determining what impact the change will have on their area. Not every Engineering Change will effect each department.

4.0 ENGINEERING CHANGE IMPLEMENTATION PROCEDURE

4.1 Issuing the Engineering Change

The Engineering Change and Agenda are prepared by the Engineering Change Analyst, and are issued to Manufacturing weekly for review.

4.2 Engineering Change Review - Manufacturing

4.2.1 Manufacturing will evaluate the Engineering Change for impact on: plant capacity, manpower, subcontractors. Manufacturing analyzes the Engineering Change for accuracy, manufacturability, solution given, cost effectiveness, and product structure. Manufacturing reviews the recommended disposition of parts. Any concerns or problems found with the Engineering Change must be discussed and resolved with the other departments prior to determining the effectivity date and prior to Engineering Change sign-off.

4.2.2 Manufacturing reviews the Item Master Database to confirm source code. Any changes required should be coordinated with Production Control and Engineering. Manufacturing assesses the impact on in-house tooling and works with the Purchasing to determine what impacts there will be on tooling at the subcontractors and vendors tooling. The estimated completion date for altering existing in-house tooling or creating new in-house tooling is determined.

4.2.3 For parts in the Engineering Change that require rework, Manufacturing must work with Purchasing and Production Control to determine the most effective way to do the rework, considering cost and manpower availability. In some instances, it might be cost effective to scrap the part. In those cases, Engineering must be requested to change the part's disposition.

4.2.4 If the Engineering Change involves changes to parts, Manufacturing must assess the impact on part storage at the work stations. The time required to set up part bins and/or part cards must be determined.

4.2.5 Manufacturing must determine if any manufacturing personnel training is required as a result of the Engineering Change.

4.3 Engineering Change Review - Production Control

4.3.1 Production Control reviews the Engineering Change for product structure and evaluates the source codes on the Item Master Database. Any changes required should be coordinated with Engineering and Manufacturing. For any parts manufactured in-house with a disposition of "rework", "obsolete", or "scrap", the planner must fill in on-hand quantities including accurate non-nettable quantities from the floor.

4.3.2 If part rework is required, Production Control works with Manufacturing to determine how long the rework will take and how efficiently it can be scheduled into production. They coordinate with the purchasing group to determine the best course to take for rework.

4.3.3 If any training of Production Control personnel is required because of an Engineering Change, the Manager must estimate the hours needed.

4.4 Engineering Change Review - Purchasing

4.4.1 Purchasing reviews the Engineering Change to see what parts are impacted. If they have any problems with the Engineering Change they must advise Engineering and resolve these issues prior to assigning an effectivity date and prior to Engineering Change sign-off. Purchasing obtains quotes from vendors for pricing of new parts.

4.4.2 Purchasing must check for open purchase orders on the parts and fill in on-hand nettable quantities and obtain part quantities from the floor. They fill in estimated month's usage using a part's Material Requirements Planning report. Purchasing must have the vendor advise the company as to work in process and finished goods liability for the part(s) at the vendor's location.

4.4.3 After consulting with the Manufacturing to see if vendor or subcontractor tooling is impacted by the Engineering Change, Purchasing must work with the vendor and/or subcontractor to determine the time required to change or make new tooling.

4.4.4 For any parts that must be reworked, Purchasing works with Manufacturing and the vendor to determine the most cost-effective way to do the rework. The vendor should be consulted to see if the best method has been found and the most cost effective means of change and/or manufacture, if so, they may be able to contribute assistance in refining or further cost reducing the Engineering Change.

4.5 Engineering Change Review - Quality

4.5.1 Quality reviews the Engineering Change for quality and inspection concerns and solution proposed. Quality also reviews the disposition of parts. Any problems or concerns with the Engineering Change must be resolved with Engineering and Manufacturing prior to determining an effectivity date and prior to Engineering Change sign-off.

4.5.2 If the Engineering Change impacts any Quality procedures, the time required to rework or write new procedures is calculated and the hours needed.

4.5.3 Quality assesses the impact the Engineering Change might have on Quality test equipment and software.

4.5.4 Any time required to train Quality personnel as a result of an Engineering Change must be determined.

4.6 Engineering Change Review - Accounting

4.6.1 Accountant reviews the Engineering Change for accounting concerns and analyzes the cost impact of parts with a disposition of obsolete or scrap. Any concerns with the disposition of parts must be discussed with Engineering and Quality and a solution arrived at prior to determining effectivity date and prior to Engineering Change sign-off.

4.6.2 Accounting considers the time involved in entering and/or changing the Item Master Database initiated by the Engineering Change. If that or any other concerns about an Engineering Change effectivity date arise, those concerns must be explained.

4.7 Engineering Change Review - Sales/Marketing

4.7.1 Sales/Marketing reviews pertinent Engineering Changes and assesses any inside sales personnel training required as a result of Engineering Change activity. They estimate the hours required for training.

4.7.2 Sales/Marketing also reviews any Item Master Database initiated by the Engineering Change so that any required changes can be made before the Engineering Change is signed-off. These forms should remain with the Engineering Change until sign-off.

4.7.3 Sales/Marketing reviews pertinent Engineering Changes for impacts on marketing literature and manuals. Any concerns or questions about the Engineering Change should be discussed with the appropriate parties prior to Engineering Change sign-off.

4.7.4 Sales/Marketing reviews the Engineering Change to assess any impact the change might have on the sales office and the customer. They also review the Item Master Database initiated by the Engineering Change so that any required changes can be made before the Engineering Change is signed-off.

4.7.5 Customer Support reviews the Engineering Changes to ensure that Customer Support requirements have been met. If there is a problem with an Engineering Change, they must get it resolved prior to Engineering Change sign-off.

4.8 Determining Engineering Change Effectivity Date

4.8.1 It is extremely important that a realistic and accurate effectivity date be assigned to every Engineering Change. Poorly assigned effectivity dates can result in obsolescence of parts that are still usable, can lead to inventory inaccuracies when using backflushing techniques to deduct inventory, and can result in an unprepared manufacturing plant.

4.8.2 The Engineering Change Analyst conducts a meeting every week, at which the Engineering Change effectivity dates are determined, one day prior to the formal Engineering Change sign-off meeting. By this time the Item Master Database must be completed, as they will be used to finish the Engineering Change Implementation documentation. At this meeting, the documentation is analyzed and based on the information listed, and on the input from the parties present an effectivity date for the Engineering Change is determined.

Title: **SUFFIX ENGINEERING CHANGE**		Number: EP-6-10 Revision: A
Prepared by:	Approved by:	

1.0 PURPOSE

The purpose of this procedure is to provide instructions, and to assign responsibilities for the preparation and release of suffix engineering changes.

2.0 APPLICABLE DOCUMENTS

None

3.0 SUFFIX ENGINEERING CHANGE PROCEDURE

3.1 General Requirements

A Suffix Engineering Change shall be generated as a supplement to a released Engineering Change when Field or Technical Manual updates are made necessary by the nature of the original Engineering Change. A Suffix Engineering Change may also be used to correct a minor error in the documentation of the original Engineering Change, according to the requirements and limitations set forth in this document. No additional change documentation outside these guidelines shall be allowed in a Suffix Engineering Change package.

4.0 SUFFIX ENGINEERING CHANGE PROCEDURE

Suffix Designation	Definitions/Requirements
EC Number + A	Reserved for field updates Field Instructions made necessary by the original Engineering Change. Suffix "A" is originated by Engineering in conduction with Field Engineering. The Change Control Board is chaired by the Manager of cognizant Engineering group. One field Suffix Engineering Change will be allowed per Engineering Change. Additional field updates are the responsibility of Field Engineering and must be coordinated through Engineering and requires a new Engineering Change.
EC Number + M	Reserved for updates to technical manuals made necessary by original Engineering Change. Suffix "M" is originated by Technical Publications and coordinated with the appropriate Engineering group to ensure correlation with the content of the original Engineering Change. The Change Control Board is chaired by the Manager of cognizant Engineering group, unless delegated by same to Technical Publications management.
EC Number + N	This suffix is used to correct minor reprographic, documentation, or non-dimensional drafting errors in original Engineering Change package which constitute non-functional changes to the content of the Engineering Change. Only those changes which require no additional Engineering Change costing may be documented in a "N" suffix. Suffix "N" is originated by the appropriate Engineering Documentation group, Drafting, or Engineering. No Change Control Board is required for release of the "N" suffix. However, authorization signature from the Manager of the cognizant Engineering Documentation group is required.

Title: **SUFFIX ENGINEERING CHANGE**	Number: EP-6-10

EC Number + P

This suffix is used to correct non-functional errors in the original Engineering Change package (i.e., omission of an affected bills of materials, down level documentation, drafting error, etc.). Functional changes to original Engineering Change shall not be documented in a Suffix Engineering Change; a new Engineering Change must be generated. Engineering originates the "P" suffix and the Change Control Board is chaired by the appropriate Engineering manager.

Note: *Any drawing affected by a "P" suffix will require the appropriate revision level change.*

EC Number + X

This suffix is used for the sole purpose of changing a disposition from the original Engineering Change. Suffix "X" is originated by Engineering and the Change Control Board is chaired by the appropriate Engineering manager.

Title: **BILL OF MATERIAL CHANGE**		Number: EP-6-11 Revision: A
Prepared by:	Approved by:	

1.0 PURPOSE

The purpose of this procedure is to provide a method for making changes to released Bills of Materials when an Engineering Change is not required. It will also provide instructions for preparing and submitting the Bill of Material Change form.

2.0 APPLICABLE DOCUMENTS

Engineering Procedure:

EP-2-2, Document Number Assignment Logbook

Engineering Form:

E040, Bill of Material Change

3.0 GENERAL REQUIREMENTS

3.1 There are two distinct types of Bills of Materials. There are Bills of Materials which are part of the Engineering structure of the product; that is, Bills of Materials which are generated and authorized by Engineering for the purpose of defining the product. Changes to these Bills of Materials must be authorized by an Engineering Change.

There is also, non-Engineering Bills of Materials for purposes such as ease of manufacturability, material direction, recondition. These Bills of Materials are not part of the Engineering structure, and therefore do not require the authorization of an Engineering Change when changes are necessary. Changes to non-Engineering Bills of Materials may be processed directly through Manufacturing. These will be handled on an individual basis.

3.2 The Bill of Material change is intended to provide a timely, low-cost method whereby changes to Bills of materials within the Engineering structure can be properly documented, authorized, and communicated. Because the Bill of Material change is actually an abbreviated form of the complete Engineering Change procedure, it only allows for usage in changes which meet the criteria specified within this document. All other Engineering Changes must be processed through the complete Engineering Change Procedure.

The complete Engineering Change procedure is required if:

a) It causes a form, fit, or function change to the assembly beyond the extent specified herein.

b) The change causes a modification on the current assembly drawing.

c) An increase in total cost greater than $.25 per product is incurred.

3.3 A Bill of Material Change can serve to exchange a part or correct a quantity if the Bill of Material does not match the drawing.

Only one assembly may be effected per a Bill of Material change.

Depending on time limitations, the originator may expedite the sign-off process by hand-carrying the document for approvals.

The change must constitute a "use-as-is" disposition.

4.0 BILL OF MATERIAL CHANGE PROCEDURE

This procedure is divided into the following parts:

- Bill of Material Change
- Bill of Material Change Form Preparation

The following procedure describes who is responsible and what they are supposed to do for each processing step.

4.1 Originator

Steps

1. Prepare the Bill of Material Change form E040 to request a change to released Bills of Materials. (See Figure 46. for an example of the form.)

2. Forward the completed Bill of Material Change form to your manager for review and approval.

4.2 Manager

3. Verify that the Bill of Material Change form is complete and correct. If approved, sign and date, then return the form to the Originator for further processing.

4.3 Originator

4. Forward the approved Bill of Material Change form to Document Control for the assignment of a Bill of Material Change number.

4.4 Document Control

5. Assign the next available number from the Document Number Assignment Logbook EP-2-2, under Paragraph 12.0, Bill of Material Number Assignment Log. Then enter the number in the Bill of Material Change form E040.

6. Forward a copy of the Bill of Material Change form to the Originator.

4.5 Originator

7. Circulate the Bill of Material Change to the responsible areas for their input.

4.6 Manufacturing Engineering

8. Review the Bill of Material Change to determine whether it is process, design, or document related. If process related, Manufacturing Engineering answers and implements a solution. Forward the Bill of Material Change form to Production Control.

4.7 Production Control

9. Complete the cost data.

10. Verify the availability of both old and new materials.

11. Set the effectivity accordingly.

12. Sign and forward the Bill of Material Change form to the Engineering.

4.8 Engineering

13. Verify that the change is valid and within the requirements as stated in this document; and will be signed as appropriate. If the request is rejected, indicate the reason and return the Bill of Material Change form to the originator to resolve the issues.

14. If accepted, sign and date the Bill of Material Change form and forward to Manufacturing.

4.9 Manufacturing

15. Verify that the change constitutes a use-as-is disposition. If rejected, indicates the reason and return it to the originator to resolve the issues.

16. If Accepted, sign and forward the Bill of Material Change form to Quality.

4.10 Quality

17. Verify that the change will not impair the quality of the product, and that with the change, the product is assurable. If rejected, indicate the reason and return the Bill of Material Change form to the originator to resolve the issues.

18. If accepted, sign and forward the Bill of Material Change form to Field Engineering.

4.11 Field Engineering

19. Review the change for field impact. If unacceptable, resolve issues with the originator.

20. If acceptable, sign the Bill of Material Change form and forward it to Production Control.

4.12 Production Control

21. Perform final verification and input into the accounting system.

22. Forward the Bill of Material Change form to Document Control.

4.13 Document Control

23. Distribute copies of the approved Bill of Material Change form to all signatories and other effected areas, per normal Engineering Change Distribution.

BILL OF MATERIAL CHANGE

Engineer:
Department No.: ❶ Date:
BOM Change No.: ❷

Manager:
Department No.: ❸ Date:
Cost: ❹ ☐ Increase ☐ Decrease

Description of Change: ❺

Disposition: Use-as-is. ❻ Effectivity Date:

Part No.: ❼ Part Name: Drawing No.:

Justification: ❽

Reject Reason: ❾

Authorizations:
Engineering: ❿ Date:

Manufacturing: Date:

Quality: Date:

Purchasing: Date:

Form E040 (Procedure EP-6-11)

Figure 46. Bill of Material Change Form

5.0 BILL OF MATERIAL CHANGE FORM PREPARATION

The procedure for processing the Bill of Material Change form shall be followed by each individual responsible for entering information on the form. Each circled number below corresponds to the circled number on the Bill of Material Change form E040. (See Figure 46. for an example of the form.)

5.1 Engineer

❶ Enter your Name, Department Number and the Date.

5.2 Document Control

❷ Assign the next available part number from the Document Number Assignment Logbook EP-2-2, under Paragraph 12.0, Bill of Material Change Number Assignment Log. Then enter the number in the Bill of Material Change form E040.

5.3 Manager

❸ Enter your Name, Department Number and Date, upon approval.

5.4 Engineer

❹ Check the appropriate box to indicate an increase or decrease in the cost of the end product.

❺ Describe the change in detail.

❻ Enter the disposition of either, use-as-is or enter an effectivity date.

❼ Enter Part number, Part Name, and Drawing Number.

❽ Enter the justification of why the Bill of Material Change is needed.

❾ Enter the reason for rejection.

❿ Each Department representative enters their Name and Date, if approved.

Title: **DOCUMENT CHANGE NOTICE**		Number: EP-6-12 Revision: A
Prepared by:	Approved by:	

1.0 PURPOSE

The purpose of this procedure is to provide instructions, and assign responsibilities for preparing and submitting the Document Change Notice form.

2.0 APPLICABLE DOCUMENTS

Engineering Procedure:

EP-2-2, Document Number Assignment Logbook

Engineering Form:

E041, Document Change Notice

3.0 GENERAL REQUIREMENTS

Any change to an Engineering Drawing may be documented on a Document Change Notice form. The form is used when an illustration of a change is necessary. Changes documented on the Document Change Notice form shall be shown in terms of the WAS and NOW conditions of a part. This form is used within the Engineering Change package, to isolate and document required changes to Engineering Drawings.

Engineering changes documented on the Document Change Notice shall be incorporated into document masters immediately upon release of the Engineering Change package.

4.0 DOCUMENT CHANGE NOTICE PROCEDURE

This procedure is divided into the following parts:

- Document Change Notice
- Document Change Notice Form Preparation

The following procedure describes who is responsible and what they are supposed to do for each processing step.

4.1 Originator

Steps

1. Prepare the Document Change Notice form E041 to document changes to Engineering Drawings. (See Figure 47. for an example of the form.)

2. Forward the completed Document Change Notice form to your manager for review and approval.

4.2 Manager

3. Verify that the Document Change Notice form is complete and correct. If approved, sign and date, then return the form to the Originator for further processing.

4.3 Originator

4. Forward the approved Document Change Notice form to Document Control for the assignment of a Document Change Notice number.

4.4 Document Control

5. Assign the next available number from the Document Number Assignment Logbook EP-2-2, under Paragraph 18.0, Document Change Notice Number Assignment Log. Then enter the number in the Document Change Notice form E041.

6. Forward a copy of the Document Change Notice form to the Originator.

DOCUMENT CHANGE NOTICE			
Originator: Department No.: ❶		Date:	Doc. Change Notice No.: ❷
Change Control Board Chairperson: Department No.: ❸		Date:	Page: ❹ of
Document Number: ❺	Description:		Revision:
Associated Engineering Change Number: ❻			
WAS Condition: ❼			
NOW Condition:			
Incorporated By: ❽		Incorporation Date:	

Form E041 (Procedure EP-6-12)

Figure 47. Document Change Notice Form

5.0 DOCUMENT CHANGE NOTICE FORM PREPARATION

The procedure for processing the Document Change Notice form shall be followed by each individual responsible for entering information on the form. Each circled number below corresponds to the circled number on the Document Change Notice form E041. (See Figure 47. for an example of the form.)

5.1 Originator

❶ Enter your Name, Department Number and the Date.

5.2 Document Control

❷ Assign the next available part number from the Document Number Assignment Logbook EP-2-2, under Paragraph 18.0, Document Change Notice Number Assignment Log. Then enter the number in the Document Change Notice form E041.

5.3 Change Control Board Chairperson

❸ Enter your Name, Department Number and Date, upon approval.

5.4 Originator

❹ Enter the number of pages. Example, Page: 1 of 3.

❺ Enter the document number, description and its revision.

❻ Enter the associated Engineering Change number.

❼ Enter an illustration of the part prior to this change in the WAS condition section, then enter the change illustration in the NOW condition section.

5.5 Engineer

❽ Enter your name and date that this change is incorporated.

Title: **SPECIFICATION CHANGE NOTICE**		Number: EP-6-13 Revision: A
Prepared by:	Approved by:	

1.0 PURPOSE

The purpose of this procedure is to provide instructions, and assign responsibilities for preparing and submitting the Specification Change Notice form.

2.0 APPLICABLE DOCUMENTS

Engineering Procedure:

EP-2-2, Document Number Assignment Logbook

Engineering Form:

E042, Specification Change Notice

3.0 GENERAL REQUIREMENTS

Any change to an Engineering Specification may be documented on a Specification Change Notice form. Changes documented on the Specification Change Notice form shall be shown in terms of the WAS and NOW conditions of a specification. This form is used within the Engineering Change package, to isolate and document required changes to Engineering Specifications.

Engineering changes documented on the Specification Change Notice shall be incorporated into document masters immediately upon release of the Engineering Change package.

4.0 SPECIFICATION CHANGE NOTICE PROCEDURE

This procedure is divided into the following parts:

- Specification Change Notice
- Specification Change Notice Form Preparation

The following procedure describes who is responsible and what they are supposed to do for each processing step.

4.1 Originator

Steps

1. Prepare the Specification Change Notice form E042 to document changes to Engineering Specifications. (See Figure 48. for an example of the form.)

2. Forward the completed Specification Change Notice form to your manager for review and approval.

4.2 Manager

3. Verify that the Specification Change Notice form is complete and correct. If approved, sign and date, then return the form to the Originator for further processing.

4.3 Originator

4. Forward the approved Specification Change Notice form to Document Control for the assignment of a Specification Change Notice number.

4.4 Document Control

5. Assign the next available number from the Document Number Assignment Logbook EP-2-2, under Paragraph 19.0, Specification Change Notice Number Assignment Log. Then enter the number in the Specification Change Notice form E042.

6. Forward a copy of the Specification Change Notice form to the Originator.

<table>
<tr><td colspan="3">SPECIFICATION CHANGE NOTICE</td></tr>
<tr><td colspan="2">Originator:
Department No.: ❶ Date:</td><td>Spc. Change Notice No.:
❷</td></tr>
<tr><td colspan="2">Change Control Board Chair Person:
Department No.: ❸ Date:</td><td>Page: ❹ of</td></tr>
<tr><td colspan="3">Specification Number: ❺ Description: Revision:</td></tr>
<tr><td colspan="3">Associated Engineering Change Number: ❻</td></tr>
<tr><td colspan="3">WAS Condition:
❼</td></tr>
<tr><td colspan="3">NOW Condition:</td></tr>
<tr><td colspan="3">Incorporated By: ❽ Incorporation Date:</td></tr>
</table>

Form E042 (Procedure EP-6-13)

Figure 48. Specification Change Notice Form

5.0 SPECIFICATION CHANGE NOTICE FORM PREPARATION

The procedure for processing the Specification Change Notice form shall be followed by each individual responsible for entering information on the form. Each circled number below corresponds to the circled number on the Specification Change Notice form E042. (See Figure 48. for an example of the form.)

5.1 Originator

❶ Enter your Name, Department Number and the Date.

5.2 Document Control

❷ Assign the next available part number from the Document Number Assignment Logbook EP-2-2, under Paragraph 19.0, Specification Change Notice Number Assignment Log. Then enter the number in the Specification Change Notice form E042.

5.3 Change Control Board Chairperson

❸ Enter your Name, Department Number and Date, upon approval.

5.4 Originator

❹ Enter the number of pages. Example, Page: 2 of 3.

❺ Enter the specification number, description and its revision.

❻ Enter the associated Engineering Change number.

❼ Enter the specification prior to this change in the WAS condition section, then enter the changed specification in the NOW condition section.

5.5 Engineer

❽ Enter your name and date that this change is incorporated.

Title: **FIELD INSTRUCTIONS**		Number: EP-6-14 Revision: A
Prepared by:	Approved by:	

1.0 PURPOSE

The purpose of this procedure is to provide the general guidelines for creating and revising Field Instructions.

2.0 APPLICABLE DOCUMENTS

None.

3.0 OVERVIEW

3.1 Field instructions comprised of Bills of Materials and rework instructions that will be used to upgrade products in the field. Product Engineering and Field Support will determine the need for Field Instructions. Revision of the Field Instructions occurs through the Engineering Change procedure.

3.2 There are three types of Field Instructions:

Mandatory - Required for all field units to achieve a specified level of performance or safety.

Optional - An enhancement to a field unit that may be ordered at the customer's discretion but which does not affect product safety.

Exceptional - Required for limited, specified number of field units.

3.3 Field Instructions may be released as part of the Original Engineering Change or within a Suffix Engineering Change. A Suffix Engineering Change is used when:

It is necessary to change the text of an instruction to comply with the intent of the original Engineering Change.

When the Field Instructions were omitted from the Original Engineering Change.

When the configuration for the Field is different from Manufacturings configuration.

3.4 Field Instructions will be included in the Suffix Engineering Change as technical documents and will be processed as technical documents after Release. The review period is from 2-5 days.

3.5 Each Field Instruction addresses one functional problem: however, more than one part or update action may be required in a single Field Instruction package as long as the intent is to correct a single problem.

4.0 RESPONSIBILITIES

4.1 Field Engineering shall originate new Field Instructions as required and provide review and final copies to effected areas. Field Engineering shall update and revise existing Field Instructions as required and shall supply the input for the Engineering Change package.

4.2 Product Engineering shall be responsible for determining the need for Field Instructions, in conjunction with Field Engineering. Product Engineering shall supply initial input for Field Instructions and shall work with Field Engineering to produce an acceptable final draft copy.

4.3 Field Engineering, shall work with Product Engineering to determine the need for Field Instructions and shall receive inputs as requested from Product Engineering.

4.4 Engineering Services, shall review and make recommendation to produce an acceptable draft copy for incorporation into the final Field Instructions.

4.5 Production Control, shall provide costing of the Field Instructions and shall input into the database the information provided by Field Engineering.

4.6 Engineering Change Analyst, shall review draft copies of the Field Instructions and make recommendations for incorporation into final documents.

5.0 FIELD INSTRUCTIONS PROCEDURE

5.1 For all Class 1 Engineering Changes the Field Instructions must be created.

5.1.1 Product Engineering supplies draft input of the Field Instructions prior to the Change Control Board to define the engineering intent of the change.

5.1.2 When the Engineering Change has completed the Change Control Board cycle, Field Engineering receives a technical release distribution package from the Engineering Change Analyst.

5.1.3 Field Engineering is responsible for generating a draft of the Field Instructions from the Engineering Change package contents and Product Engineering input.

5.1.4 The Field Instructions are reviewed by the Engineering Change Analyst, and Product Engineering.

5.1.5 Field Engineering incorporates comments from these areas and distributes final copies of the Field Instructions to the Engineering Change Analyst, Production Control and Product Engineering.

5.1.6 The master of the Field Instructions are forwarded to Document Control for filing.

5.2 For all class 2 Engineering Changes it may be necessary to update an existing Field Instruction.

1. Compatible changes are in an existing Field Bill of Material.

2. Changes to Field Instructions after the Engineering Change has been closed: use a Suffix Engineering Change to make the correction. Change the Field Instruction number and replace the previous instruction on the Field Bill of Material with the new document.

5.3 If the Engineering Change has not been closed, correct the master document. No change to part number, revision, or Engineering Change is necessary.

Title: **EQUIVALENT ITEM AUTHORIZATION**		Number: EP-6-15 Revision: A
Prepared by:	Approved by:	

1.0 PURPOSE

The purpose of this procedure is to define the process, responsibilities, and requirements for the generation and release of the Equivalent Item Authorization form.

2.0 APPLICABLE DOCUMENTS

Engineering Procedure:

EP-2-2, Document Number Assignment Logbook

Engineering Form:

E043, Equivalent Item Authorization

3.0 EQUIVALENT ITEM AUTHORIZATION PROCEDURE

This procedure is divided into the following parts:

- Equivalent Item Authorization
- Equivalent Item Authorization Form Preparation

The following procedure describes who is responsible and what they are supposed to do for each processing step.

3.1 Originator

Steps

1. Prepare the Equivalent Item Authorization form E043 to request the use of a substitute equivalent item in lieu of the specified item. (See Figure 49. for an example of the form.)

2. Forward the completed Equivalent Item Authorization form to your manager for review and approval.

3.2 Manager

3. Verify that the Equivalent Item Authorization form is complete and correct. If approved, sign and date, then return the form to the Originator for further processing.

3.3 Originator

4. Forward the approved Equivalent Item Authorization form to Document Control for the assignment of an Equivalent Item Authorization number.

3.4 Document Control

5. Assign the next available number from the Document Number Assignment Logbook EP-2-2, under Paragraph 20.0, Equivalent Item Authorization Number Assignment Log. Then enter the number in the Equivalent Item Authorization form E043.

6. Forward a copy of the Equivalent Item Authorization form to the Originator for further processing.

EQUIVALENT ITEM AUTHORIZATION	
Originator: Department No.: ❶ Date:	EIA No.: ❷
Manager: Department No.: ❸ Date:	Program No.: ❹
❺ Drawing No.: ____________ Description: ____________________ Revision: ______ Change Part No.: __________ Description: ____________________ To Part No.: ____________ Description: ____________________	
❻ Effective for Serial Number ______________________________ Only. or From Serial Number ____________ Through ____________	
Purpose of the Substitution: ❼	
Authorizations: Engineering: ❽ Date:	
Manufacturing: Date:	
Quality: Date:	

Form E043 (Procedure EP-6-15)

Figure 49. Equivalent Item Authorization Form

4.0 EQUIVALENT ITEM AUTHORIZATION FORM PREPARATION

The procedure for processing the Equivalent Item Authorization form shall be followed by each individual responsible for entering information on the form. Each circled number below corresponds to the circled number on the Equivalent Item Authorization form E043. (See Figure 49. for an example of the form.)

4.1 Originator

❶ Enter your Name, Department Number and the Date when the Equivalent Item Authorization is completed.

4.2 Document Control

❷ Assign the next available part number from the Document Number Assignment Logbook EP-2-2, under Paragraph 20.0, Equivalent Item Authorization Number Assignment Log. Then enter the number in the Equivalent Item Authorization form E043.

4.3 Manager

❸ Sign and Date the Equivalent Item Authorization form, upon approval.

4.4 Originator

❹ Enter the program name.

❺ Enter the drawing number, description and revision letter. Enter the part number and description of the specified part and the part number of the recommended substitute part.

❻ Enter the limitation as to the serial number effectivity for which the substitution would be permissible.

❼ Enter the purpose of the substitution.

4.5 All Departments

❽ Enter the signature of the responsible Engineer, Manufacturing and Quality personnel and the date the document was approved.

SECTION 7

DOCUMENT CONTROL

Title: **DOCUMENT CONTROL SYSTEM**		Number: EP-7-1 Revision: A
Prepared by:	Approved by:	

1.0 PURPOSE

The purpose of this procedure is to define the requirements for the establishment and maintenance of engineering project files. The files will be used to provide an accessible permanent record of design decisions and development history for all engineering projects.

2.0 APPLICABLE DOCUMENTS

None.

3.0 PROCEDURE

3.1 General

Documentation is the only reliable way of passing information to other people. Spoken, non-documented communication is highly effective, fast, and a bi-directional means of transferring information. Undocumented information, however, often tends to be forgotten in the long run. Hence, every noteworthy piece of information must be clearly documented. The following words can be used for describing the basic principle of documentation: "The job isn't finished until the paperwork is done."

3.2 Product development activity involves mainly two types of documents: 1) design documents, 2) production documents.

3.3 Immediately after the product development decision, a development file is established for the design project. All documents produced in conjunction with the product development are assembled in this file, maintained as long as the product exists. For large systems, several files should be established: one for system level and others for the different units. As far as the Engineer is concerned, the development file is a work book in which all daily notes, design, production, and service documents are assembled. The Development File can also be used for monitoring the progress of the product development project. There are eight chapters in the development file, and it resides in Document Control.

3.4 The development file is established by the project leader who will provide a title page and list of contents. The file is continuously updated. All important pieces of paper, even sketches, scrap paper or notes will be filed in the appropriate location. The reason for a decision must be traceable. All pieces of paper must be provided with date and initials.

3.5 The Requirements Specification that is written during the product development planning stage is also placed in the development file. In the Requirement Specification, all manufacturing and service documents made during the development project are specified. These manufacturing and service documents are usually produced in accordance with standard operating procedures.

3.6 The number of documents in the development file increases and their quality improves during the product development procedure. At the Preliminary Design Review, the preliminary manufacturing and service documents are presented to the participants. Those agreed upon are accepted and handed over to the Manufacturing and Marketing Departments at the Final Design Review.

4.0 HARDWARE DOCUMENTATION

4.1 Documents During the Design Phase

The responsible Engineer shall collect and maintain the following documents:

General requirements and the objective of the design work;

Design and design review schedules;

Technical requirements;

System design; comparison of possible alternatives;

Circuit diagrams of prototypes;

Prototype measurements;

Prototype operation description;

Testing instructions;

Test equipment and connecting diagrams;

Service and maintenance instructions;

4.2 Documents for Production

During the Critical Design Review, Engineering shall hand over to Manufacturing the following documents:

Specifications: Unit and module specifications; Prototype performance; and Reliability analysis.

Manufacturing drawing sets: Master drawings list; Block diagram; Circuit diagram; Installation and assembly drawings; Wiring drawings and/or tables; Drawings for printed circuit board; Type and construction layouts; Mechanical construction drawings; Preliminary Bills of Materials.

Manufacturing instructions: Test instructions; Description of test system, test equipment, and test jigs; Work instructions for printed circuit boards; Specifications for materials; Work instructions for special processes.

4.3 Software Documentation

The responsible engineer shall collect and maintain the following documents:

Documents: Program general description; Software test descriptions and operating instructions for production; Description of physical environment; Description of program environment; System data files; Description of program modules; Program (module) generation guide; Operating instructions; Source listings; Flow charts; Program structure diagram; EPROM programming guide; Program original and safe copy (on two magnetic tapes); Program work copies, list copies, load copies, and back-up copies as required during design, testing, and system support phases.

5.0 CONTENTS OF THE DEVELOPMENT FILE

1.0 STATE OF DEVELOPMENT

1.1 Initial instructions and information given by the Engineering Manager.

1.2 Results of pre-development and other background information.

2.0 DESIGN DOCUMENTS

2.1 Production Design Plan Documents

2.1.1 Requirement Specification
2.1.2 Resources Studies
2.1.3 Organization Studies
2.1.4 Schedule
2.1.5 Product Quality Assurance Plan
2.1.6 Cost Estimate
2.1.7 Alternatives Study

2.2 Documents for Product Design Supervision

2.2.1 Purpose of Project
2.2.2 Checklists and monitoring state of maturity
2.2.3 Monitoring of expenditures
2.2.4 Calls for design reviews, minutes of design reviews, etc.

2.3 Documents and Notes for Detailed Design

2.4 Documents Concerning Quality Verification

2.4.1 Hardware
2.4.2 Software
2.4.3 Product/System entity

3.0 MANUFACTURING DOCUMENTS

3.1 System Documents

3.1.1 System Description
3.1.2 System Generation Breakdown Drawing
3.1.3 Other System Documents

3.2 Hardware Documents

3.2.1 Technical Specifications
3.2.2 System Generation Breakdown Drawing
3.2.3 Items Lists
3.2.4 Bills of Materials
3.2.5 Components and Materials Specifications
3.2.6 Assembly Drawings
3.2.7 Block Diagrams
3.2.8 Schematic Diagrams
3.2.9 Component Placement Drawings
3.2.10 PCB drawings
- PCB master drawings
- Silk screens
- Solder mask
- Films
- Fabrication drawings

3.2.11 Wiring diagrams
3.2.12 Wiring tables
3.2.13 Wiring harness drawings
3.2.14 Parts drawings
2.2.15 Label and plate drawings (also warning labels)
- Originals
- Films

3.2.16 Inductor drawings
3.2.17 Transformer drawings
3.2.18 Testing instructions
3.2.19 Sketches of jigs and tools
3.2.20 Other hardware documents

3.3 Software Documents

3.3.1 Software descriptions
3.3.2 Technical specifications
3.3.3 Codes
3.3.4 Software label data
3.3.5 Other software documents

4.0 SERVICE DOCUMENTS

4.1 System Program or Operation Manual
4.2 Technical Manual
4.3 Installation and Service Manual
4.4 Other Service Documents

5.0 CUSTOMER COMPLAINTS AND OTHER PROBLEMS

6.0 DESIGN CHANGES (Engineering Changes, etc.)

7.0 OTHER DOCUMENTS

8.0 DESIGN STANDARDS

Title: **DOCUMENT RELEASE**		Number: EP-7-2 Revision: A
Prepared by:	Approved by:	

1.0 PURPOSE

The purpose of this procedure is to provide instructions, and to assign responsibilities for controlling the release of product documentation. It will also provide the instructions for preparing and submitting the Document Release Notice form.

2.0 APPLICABLE DOCUMENTS

Engineering Procedures:

EP-2-2, Document Number Assignment Logbook

EP-1-6, Product Phases

Engineering Form:

E044, Document Release Notice

3.0 DEFINITIONS

3.1 Product Documentation

Documents that are prepared and released by Engineering for the purpose of defining and controlling the manufacture and inspection of company products. The primary documents that will be controlled are: ▪ Drawings ▪ Specifications

3.2 Release

The process of transferring custody of product documentation from the originator to Document Control.

3.3 Document Release Notice Form

The Document Release Notice authorizes formal release and distribution of new and revised documents. Processed Document Release Notice forms are available for review in Document Control.

3.4 Control of Documents

After documents are released, the originator relinquishes physical control and tracking responsibilities by submitting them to Document Control for recording, copying, distributing, and filing.

3.5 Distribution

Copies of documents bearing the official stamp that identifies them as having been released. As revisions to documents are released, they will receive the same distribution as their last distribution, unless requested otherwise.

4.0 DOCUMENT RELEASE PROCEDURE

This procedure is divided into the following parts:

- Document Release Notice Procedure
- Document Release Notice Form Preparation

The following procedure describes who is responsible and what they are supposed to do for each processing step.

4.1 Engineer

Steps

1. Prepare the Document Release Notice form E044 to release new or revised documents into the documentation system. (See Figure 50. for an example of the form.)

2. Forward the completed Document Release Notice form to the Engineering Manager for review and approval.

4.2 Engineering Manager

3. Verify that the Document Release Notice form is complete and correct. If approved, sign and date, then return the form to the Engineer for further processing.

4.3 Engineer

4. Attach document originals to the Document Release Notice form then forward the package to Document Control for further processing.

4.4 Document Control

5. Enter the next available number from the Document Number Assignment Logbook EP-2-2, under Paragraph 3.0 Document Release Notice Number Assignment Log. Then enter the number in the Document Release Notice form E044.

6. Log document information from the Document Release Notice form into the Item Master Database.

7. Run copies of the documents that are attached to the Document Release Notice form and red stamp them with the appropriate stamp and date. The stamp is determined by the type of release: Development, Preproduction or Production.

8. Sign and date the form. This date is the official release of documents into the system.

9. Forward a copy of the Document Release Notice form to the Engineer.

10. Attach copies of released documents to the Document Release Notice form and forward it to either a or b:

    a. **Production Control** for distribution to Manufacturing. Down level copies of documents shall be returned to Document Control to be destroyed.

b. **Purchasing** for distribution to the vendors, as necessary. Purchasing is responsible for notifying vendors about destroying copies of superseded documents.

11. File the completed Document Release Notice form.

12. File the new or revised document originals that were attached to the Document Release Notice form.

13. Red stamp "OBSOLETE" on the down level originals that are on file, then date.

14. Destroy all copies of the superseded documents that are on file throughout the company.

DOCUMENT RELEASE NOTICE				
Engineer: ❶		Date:		Document Release Notice No: ❷
Eng. Manager: ❸		Date:		Priority: ❹ Routine ❑ Urgent ❑
❺ Document Release:	Development ❑	Preproduction ❑	Production ❑	

DOCUMENTS TO BE RELEASED				
❻ Document No.	Rev.	Document Title	Size	#Pages
Document Control: ❼			Release Date:	

Form E044 (Procedure EP-7-2)

Figure 50. Document Release Notice Form

5.0 DOCUMENT RELEASE NOTICE FORM PREPARATION

The procedure for processing the Document Release Notice form shall be followed by each individual responsible for entering information on the form. Each circled number below corresponds to the circled number on the Document Release Notice form E044. (See Figure 50. for an example of the form.)

5.1 Engineer

❶ Enter your name and date when the form is complete and ready for processing.

5.2 Document Control

❷ Enter the next available sequential number from the Document Release Notice Number Assignment Logbook EP-2-2 under Paragraph 3.0 Document Release Notice Number Assignment Log.

5.3 Engineering Manager

❸ Enter your name and date, if approved.

5.4 Engineer

❹ Mark the appropriate box:

Routine - will be processed within five days.

Urgent - Document Release Notice forms will be processed the same day they are received.

❺ Mark the appropriate box. This is determined by the type of release. (Ref. EP-1-6, Product Phases.)

▪ Development ▪ Preproduction ▪ Production

❻ Enter document number of each document being released. Attach a separate sheet, if releasing more documents than were allowed for on the Document Release Notice form.

Enter the revision letter of each document being released, such as, "A", "B", etc. The release process does not cause the documents revision letter to advance.

Enter the title of each document being released.

Enter the letter size of each document being released, such as, A = (8 1/2" X 11"), B = (11" X 17"), C = (17" X 24"), or D = (24" X 36").

Enter the number of pages for each document that is being released.

5.5 Document Control

❼ Sign and date the Document Release Notice form. This is the official release date of the documentation into the system.

Title: **DOCUMENT DISTRIBUTION**		Number: EP-7-3 Revision: A
Prepared by:	Approved by:	

1.0 PURPOSE

The purpose of this procedure is to define the methods used to control the distribution, and storage of product documentation.

2.0 APPLICABLE DOCUMENTS

None.

3.0 DOCUMENT DISTRIBUTION PROCEDURE

3.1 Production Release

Once the Document Release Notice has been processed, Document Control red stamps copies of the released documents (See Figure 51.) to identify them as being released, then enters the date. Released documents are automatically updated when a revision is made.

DEVELOPMENT

PREPRODUCTION

PRODUCTION

Figure 51. Development, Preproduction and Production Release Stamps

Note: *Only documents bearing the red "Production" stamp and date are authorized for use in manufacturing.*

3.2 Document Distribution

Copies of released documents will be distributed to the following as listed on the Document Release Notice form:

1. Production Control for distribution to Manufacturing.

2. Purchasing for distribution to suppliers or vendors.

3. Contract Administration for distribution to customers.

4. Document Control for filing.

3.3 Obsolete Copies

Obsolete copies of documents shall be removed from use upon the conclusion of their need by Production Control and returned to Document Control to be distorted.

3.4 Obsolete Originals

Document Control will stamp the obsolete document originals (See Figure 52.) to identify them as being obsolete, then enter date in the box.

OBSOLETE 10/19/99

Figure 52. Obsolete Original Stamp

3.5 Quotation Copies

To request copies of documents for Quotation purposes, prepare the Copy Request form. The documents will be red stamped (See Figure 53.) to identify them as Quotation copies. These copies will not receive an automatic distribution of revisions.

QUOTATION

Figure 53. Quotation Copy Stamp

3.6 Personal Copy

Anyone may request copies of released documents by submitting a Copy Request form. The documents will be red stamped (See Figure 54.) to identify them as not being maintained. These copies will not receive an automatic distribution of revisions.

WILL NOT BE MAINTAINED

Figure 54. Personal Copy Stamp

3.7 Copy Check-out

Anyone may check out a copies of released documents by preparing an "OUT CARD" in Document Control. Copies that have been checked out shall to be returned after use.

3.8 Document History File

A complete history file is maintained for all revisions of released document originals in Document Control.

Title: **ENGINEERING CHANGE/PART NUMBER INPUT**		Number: EP-7-4 Revision: A
Prepared by:	Approved by:	

1.0 PURPOSE

The purpose of this procedure is to give instructions for the preparation of the Engineering Change/Part Number cross reference file.

2.0 APPLICABLE DOCUMENTS

Engineering Form:

E045, Engineering Change/Part Number History Input

3.0 ENGINEERING CHANGE/PART NUMBER INPUT PROCEDURE

The Engineering Change/Part Number Input is a system used to keep track of and report past and future changes of engineering designs.

This procedure is divided into the following parts:

- Engineering Change/Part Number
- Engineering Change/Part Number Input Form Preparation

Title: **ENGINEERING CHANGE/PART NUMBER INPUT**	Number: EP-7-4

The following procedure describes who is responsible and what they are supposed to do for each processing step.

3.1 Originator

Steps

1. Prepare the Engineering Change/Part Number Input form E045 to establish a cross reference file of a changed part and the Engineering Change that changed the part. (See Figure 55. for an example of the form.)

2. Forward the completed form to Document Control for filing.

ENGINEERING CHANGE/PART NUMBER INPUT

❶

Document Number: ________________ Title: ____________________________________

Number of Sheets: ___ Size: ___ Bill of Material Sheets: ___

Rev.	Eng. Change No.	Pending Date:	CCB Status Accept	Reject	Eng Change Release Date:	Incroporation Date:	Note/Comments:
❷	❸	❹	❺		❻	❼	❽

Form E045 (Procedure EP-7-4)

Figure 55. Engineering Change/Part Number Input Form

4.0 ENGINEERING CHANGE/PART NUMBER INPUT FORM PREPARATION

The procedure for processing the Engineering Change/Part Number Input form shall be followed by each individual responsible for entering information on the form. Each circled number below corresponds to the circled number on the Engineering Change/Part Number Input form E045. (See Figure 55. for an example of the form.)

4.1 Originator

❶ Enter the Document number and Title, number of sheets, document size, Bill of Material sheets.

❷ Enter the current revision letter for the document.

❸ Enter the Engineering Change number that was issued against the document.

❹ Enter the date when the Engineering Change is first received in Document Control.

❺ If the Engineering Change is approved by the Change Control Board, enter the date of the approval in the Accept column. If the Engineering Change is rejected by the Change Control Board, enter the date in the Reject column.

❻ Enter the date when the Engineering Change status has been entered into the appropriate Item Master database.

❼ Enter the date when the changes from the Engineering Change have been incorporated into the original documents.

❽ Enter any notes or comments pertaining to the Engineering Change or the document.

Title: **SECURITY MARKING OF DOCUMENTS**		Number: EP-7-5 Revision: A
Prepared by:	Approved by:	

1.0 PURPOSE

The purpose of this procedure is to define the responsibilities for the handling and protection of confidential information. It also contains a listing of the types of information to be considered confidential.

2.0 APPLICABLE DOCUMENTS

None.

3.0 OVERVIEW

There are thousands of things known to the company which are not known by competitors. The competitors would be strengthened by learning what the company knows. The company must protect sensitive information and keep it out of the hands of competitors.

4.0 RESTRICTED AND CONFIDENTIAL INFORMATION

4.1 The information which may be of value to competitors or could damage the business, such as, financial or market position must be clearly marked "Confidential", and be kept in a locked cabinet, safe or drawer when not being used. Such information must be hand carried or put in the Company mail distribution in Confidential envelopes, and must be disposed of in one of the confidential trash receptacles to be destroyed.

4.2 Sensitive information consists of virtually everything related to the business, such as:

Names of customers
Names of suppliers
Names of vendors
Everything related to proprietary technology
Markets the company is pursuing
New products in development
Future plans of the company
Everything concerning operations and methods
Projections of future performance
Profit/loss statements and supporting schedules
Business plans and budgets
Salary planning

5.0 CONFIDENTIAL INFORMATION

5.1 The information which may not necessarily cause damage to the company business, financial or market position, but which could cause some difficulty or inconvenience to the company or employees.

5.2 This material must be clearly marked "Confidential", and kept out of sight when not in use. It may be sent through in-house mail in normal reusable envelopes, and can be disposed of in any regular trash receptacle, provided it has been torn up in advance.

5.3 Examples of information in this category include, but are not limited to:

Organizational listings

Home phone numbers

5.4 Access to "Restricted and Confidential" material is limited to those employees who are authorized to know. Access to "Confidential" material is limited to the appropriate employees.

6.0 SHREDDING OF DISCARDED INFORMATION

6.1 There are two types of disposal containers. One type is for sensitive information and the other is for trash. The only thing that should go in the trash container is trash, such as chewing gum, wrappers, pop cans, magazines, junk mail, etc. Everything that has anything to do with the company is to be placed in the other container for destruction. Use your judgement, but always ask yourself if the information you are discarding could be used by the competitor before you discard it. Put it into a container for shredding unless you are sure that it would not be helpful to the competitor.

> **Note:** *Tearing a document into pieces and then discarding it into a trash container is not an acceptable alternative to shredding.*

7.0 VERBAL INFORMATION

7.1 Do not tell anyone anything about the company that they do not have a need to know. Think about how much can be learned about competitors by talking to customer's suppliers, and vendors, then ask yourself what competitors are learning about the company from these sources.

7.2 Do not talk about a subject just because you are familiar with it. Information is power, use it with care.

8.0 RESPONSIBILITIES

8.1 All employers are responsible for taking adequate steps to ensure their discretion and that of their employees when dealing with confidential information. This responsibility includes handling, storing and destroying such information in accordance with the rules set forth in the categories above.

8.2 Security is responsible for removing material from the confidential trash collection boxes and ensuring that it is destroyed.

Title: **COPY REQUEST**		Number: EP-7-6 Revision: A
Prepared by:	Approved by:	

1.0 PURPOSE

The purpose of this procedure is to define the method used for requesting copies of product documentation that is maintained by Document Control.

2.0 APPLICABLE DOCUMENTS

Engineering Form:

E046, Copy Request

3.0 DEFINITION

3.1 Product Documentation

Plans
Reports
Procedures
Specifications
Drawings
Bills of Materials

4.0 COPY REQUEST PROCEDURE

This procedure is divided into the following parts:

- Copy Request
- Copy Request Form Preparation

The following procedure describes who is responsible and what they are supposed to do for each processing step.

4.1 Originator

Steps

1. Prepare the Copy Request form E046 to request copies of product documentation. (See Figure 56. for an example of the form.)

2. Forward the completed Copy Request form to Document Control for processing.

4.2 Document Control

3. Run copies, stamp and forward copies to the originator.

COPY REQUEST

❶

Originator: ______________________ Department No.: _____________ Date: __________

Date Required: __________ ❷ ❑ Pick-up ❑ Mail

Document No.	Rev.	Description	Sheets Qty.	No. of Copies
❸	❹	❺	❻	❼

Form E046 (Procedure EP-7-6)

Figure 56. Copy Request Form

5.0 COPY REQUEST FORM PREPARATION

The procedure for processing the Copy Request form shall be followed by the individual responsible for entering information on the form. Each circled number below corresponds to the circled number on the Copy Request form E046. (See Figure 56. for an example of the form.)

5.1 Originator

❶ Enter your Name, Department Number and the Date when the Copy Request is completed.

❷ Enter the date that the copies are required. Check the appropriate box depending on if you are going to pick-up your copies or if you want them mailed.

❸ Enter the document number(s) that you are requesting copies of.

❹ Enter the revision of the document(s).

❺ Enter the description of the document(s).

❻ Enter the number of sheets of this document(s).

❼ Enter the number of copies that you are requesting.

Title: **DOCUMENT ORIGINAL WITHDRAWAL**		Number: EP-7-7 Revision: A
Prepared by:	Approved by:	

1.0 PURPOSE

The purpose of this procedure is to define the method used for withdrawing document originals from Document Control to incorporate changes.

2.0 APPLICABLE DOCUMENTS

Engineering Form:

E047, Document Original Withdrawal

3.0 COPY REQUEST PROCEDURE

This procedure is divided into the following parts:

- Document Original Withdrawal
- Document Original Withdrawal Form Preparation

The following procedure describes who is responsible and what they are supposed to do for each processing step.

Title: **DOCUMENT ORIGINAL WITHDRAWAL**	Number: EP-7-7

3.1 Originator

Steps

1. Prepare the Document Original Withdrawal form E047 to withdraw document originals from Document Control. (See Figure 57. for an example of the form.)

2. Forward the completed Document Original Withdrawal form to Document Control for processing.

3.2 Document Control

3. Pull the original document and forward it to the originator.

DOCUMENT ORIGINAL WITHDRAWAL	
Originator: Department No.: ❶ Date:	EC Number: ❷
Manager: Department No.: ❸ Date:	Bill of Material: ❑ Yes ❑ No ❹
(Only one original per request.) ❺ Drawing No.: ____________ Revision: _______ Sheet Size _______	
Reason: ❻	

Form E047 (Procedure EP-7-7)

Figure 57. Document Original Withdrawal Form

4.0 DOCUMENT ORIGINAL WITHDRAWAL FORM PREPARATION

The procedure for processing the Document Original Withdrawal form shall be followed by the individual responsible for entering information on the form. Each circled number below corresponds to the circled number on the Document Original Withdrawal form E047. (See Figure 57. for an example of the form.)

4.1 Originator

❶ Enter your Name, Department Number and the Date when the Document Original Withdrawal form is completed.

❷ Enter the associated Engineering Change number.

4.2 Manager

❸ Sign and Date the Document Original Withdrawal form, upon approval.

4.2 Originator

❹ Check the appropriate Bill of Material box.

❺ Enter the drawing number, revision and the sheet size.

❻ Enter reason for request.

Title: **LITERATURE SEARCH/ACQUISITION**		Number: EP-7-8 Revision: A
Prepared by:	Approved by:	

1.0 PURPOSE

The purpose of this procedure is to explain the methods used to obtain literature from outside of the company.

2.0 APPLICABLE DOCUMENTS

None.

3.0 PROCEDURE

3.1 Numerous industry and government documents are filed in Document Control. Such documents include: papers, reports, specs, MIL-STDS, articles, journals, manuals, books, periodicals, catalogs, handbooks, and guides, etc.

Examples:

MIL-STD-271E, Nondestructive Testing Requirements for Metals

SAE AMS 4046D, Aluminum Alloy Sheet and Plate, Alclad One Side

ANSI Y14.36, Surface Texture Symbols

3.1.1 If the document you need is not on file in Document Control, you may have to search for it from outside sources.

3.2 Following are some of the sources where the above mentioned materials can be obtained.

3.2.1 PUBLIC LIBRARY has most of these documents on microfilm. First call to see if they have the document you are looking for. If they do you can go to the library and copy it.

3.2.2 UNIVERSITY LIBRARY carries government publications. Call to see if they have the document you are looking for, as their listings are limited. If they have the document, you may go to the library and copy it.

3.2.3 The UNIVERSITY ENGINEERING LIBRARY carries some industry publications. Call to see if they have the document you are looking for, as their listings are limited. If they have the document, you may go to the library and copy it.

3.3 If a document cannot be found locally, prepare a purchase requisition and go through Purchasing to obtain a copy from the following companies.

3.3.1 DOCUMENT ENGINEERING CO. has all of the documents listed in the DOD Index and will ship UPS the same day. There is a $.75 handling fee, plus postage and the price of the document. Rush orders, Next Day Air and facsimile service are also available.

Document Engineering Company, Inc.
15210 Stagg Street
Van Nuyes, California 90405
1-800 Mil Spec

3.3.2 GLOBAL ENGINEERING DOCUMENTS has the most extensive list of documents available, but their prices are generally two to three times higher than Document Engineering Co. They charge a $5 handling fee, a $10 rush fee, plus shipping costs and the price of the document.

Global Engineering Documents
2625 Hickory St. PO Box 2504
Santa Ana, California 92707-3783
1-800-854-7179

Title: **FORMS CONTROL**		Number: EP-7-9 Revision: A
Prepared by:	Approved by:	

1.0 PURPOSE

The purpose of this procedure is to establish the guidelines for forms control, including the methods for originating, ordering, and maintaining forms.

2.0 APPLICABLE DOCUMENTS

None.

3.0 OVERVIEW

All forms will be processed through Forms Control to achieve a minimum number of non-redundant, functional forms that improve the flow of information. All forms will have a form number for identification and reference. Orders for forms will be routed through Forms Control to assure the cost effective printing of current versions at the required quality level.

4.0 DEFINITIONS

4.1 Form

Any document, printed or otherwise reproduced, with spaces for the insertion of data by hand or machine. Also included are specialty labels, tags, and envelopes.

4.2 Company Forms

Any form that meets one or more of the following criteria:

Has external or inter-departmental distribution.

Has more then one ply (is multi-part), and/or

Must be printed at an outside vendor.

5.0 RESPONSIBILITIES

5.1 Forms Control

Forms Control is responsible for administration of the forms management program. The following services are included:

1. Review of all new or revised forms.
2. Assignment of form numbers for identification and reference.
3. Assistance in design and construction of forms to achieve the best type of form for a particular use.
4. Maintenance of historical usage data, masters for printing, and other documentation as required.
5. Preparation, distribution and maintenance of the forms catalog.
6. Scheduling of periodic reviews of all forms to determine obsolescence or revisions required to meet current needs.

7. Coordination of forms development activities, to ensure continuity of related forms and elimination or combination of similar forms.

8. Approval and preparation of specifications for printing.

6.0 PURCHASING

1. Place orders for outside printing and takes bids as required. Cross-check master, specifications and Purchase Order to ensure integrity of each printing job.

2. Negotiate with vendors on terms, delivery and inspection; coordinates internal delivery with personnel.

3. Maintain a list of accepted vendors and average cost for services, and provides Forms Control with a copy of this list.

4. Assure the return of printing specifications to Forms Control with complete job cost and vendor information.

5. Assure prompt return of all artwork to Forms Control upon completion of printing, with copies of the printed form for history files.

7.0 REPROGRAPHICS

1. Reproduce forms only from masters obtained from Forms Control.

2. Assure that reproduced forms meets the required standards of quality.

3. Assure return of all artwork to Forms Control upon completion of reproduction, with job cost information and two copies of the form for history.

8.0 OFFICE SUPPLIES AND EQUIPMENT

1. Stock company-wide general usage forms, and high-volume limited usage forms, only as approved by Forms Control; distribute in accordance with standard practices.
2. Provide usage information to Forms Control as requested.

9.0 END USER

1. Notify Forms Control of any forms that become obsolete or require revision.
2. When requesting design or revision of forms, provide Forms Control with the following information:
   a. The purpose of the form and the information it should provide.
   b. The usage characteristics, including where it is used and the general processing flow.
   c. The sequence of entries and any special processing requirements.
3. When requesting a form, allow sufficient time for design and printing in accordance with the following minimum time requirements:
   a. Single-part flat sheets - 10 working days.
   b. "No carbon required" forms (up to 5 parts) - 10 days to 3 weeks.
   c. Snap-out and continuous forms, tab cards, speciality labels and tags - 30 to 90 days.

10.0 FORMS CONTROL PROCEDURE

The following procedure describes who is responsible and what they are supposed to do for each processing step.

10.1 Originator

Steps

1. Forward all new forms to Forms Control for processing.

10.2 Forms Control

2. Review the new form and compare to existing forms to identify related or duplicate forms. Evaluate the form according to urgency and time required for action.

3. Contract Originator within five working days of receipt of request. If accepted, schedule for design and discuss specifications and in-dept usage information with Originator and Purchasing. Decision is made as to type of printing required and where form is to be stocked. Assign form number, and coordinate preparation of master.

4. Upon completion of design, forward proof copies and required documentation to requestor and to other personnel who will be involved with the processing of the form. Notify Originator whether initial ordering of forms will require a print request or a purchase order for outside printing.

10.3 Originator

5. Return proofs, approved or with changes noted, to Forms Control within five working days of receipt. If required, complete Purchase Order, and obtain approval. Submit print request or signed purchase order to Forms Control with approved proofs.

10.4 Forms Control

6. Correct proofs as required and prepare printing specifications. Obtain printing either by:

   a. Submit specifications and print request to in-house reprographics service, or

   b. Sending specifications, with approved Purchase Order, to Purchasing. Ensure that the Purchase Order specifies the person to whom the forms are to be delivered.

7. Establish a control file of all related documentation, and logs completed form information. Update forms log as required and periodically issues revisions to all log holders. Maintain catalog index to specify manner of obtaining forms, based on initial decision as to type of printing required.

10.5 Reprographics

8. Print form in accordance with specifications.

   Return master, two sample copies and job cost information to Forms Control; distribute remaining copies of form to the Originator, as indicated on the print request.

   or

10.6 Purchasing

9. Place order in accordance with specifications and vendor capabilities, taking bids if required. Ensure the Purchase Order clearly specifies person and department to whom forms are to be delivered. Notifies Forms Control of job cost, vendor, and expected due date.

   Return master and one printed copy to Forms Control upon receipt from vendor.

10.7 Receiving

10. Forward any forms received from vendors to requestor, as indicated on the Purchase Order.

INDEX

Printed and bound by CPI Group (UK) Ltd, Croydon, CR0 4YY
08/05/2025
01864911-0001